Historic Canals and Waterways of South Carolina

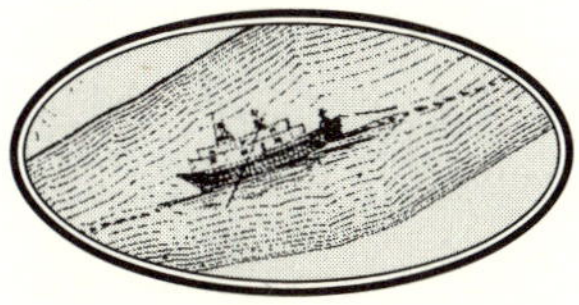

Ex. Doc. No 60 3rd Sess 46th Cong

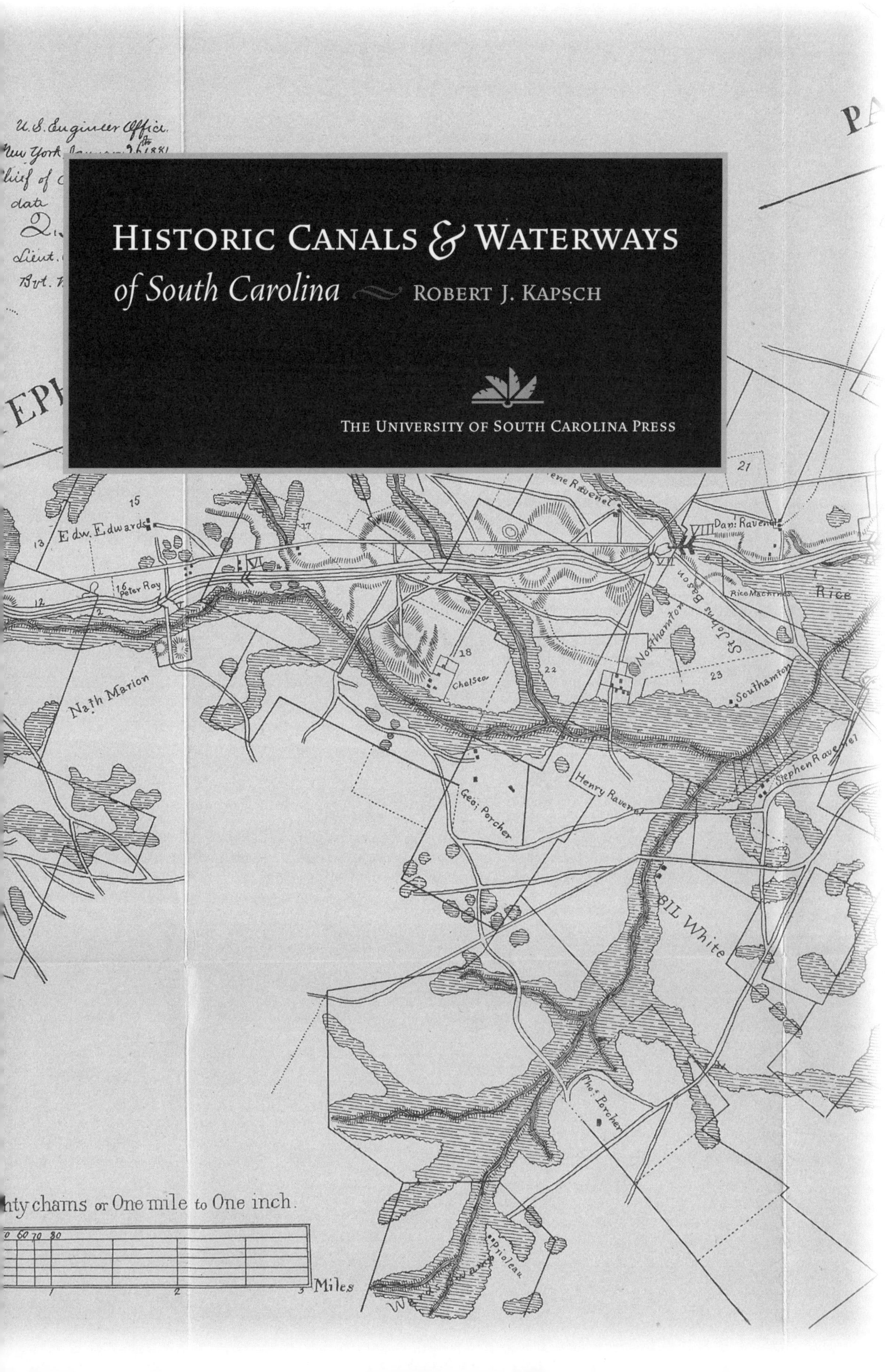

Historic Canals & Waterways of South Carolina

Robert J. Kapsch

The University of South Carolina Press

Published by the University of South Carolina Press
Columbia, South Carolina 29208

www.sc.edu/uscpress

Manufactured in the United States of America

19 18 17 16 15 14 13 12 11 10
10 9 8 7 6 5 4 3 2 1

LIBRARY OF CONGRESS CATALOGING-IN-PUBLICATION DATA

Kapsch, Robert J.
Historic canals and waterways of South Carolina / Robert J. Kapsch.
p. cm.
Includes bibliographical references and index.
ISBN 978-1-57003-867-9 (cloth : alk. paper)
1. Canals—South Carolina—History. 2. Inland water transportation—South Carolina—History. 3. Historic sites—South Carolina. I. Title.
TC624.S6K37 2010
386'.409757—dc22
2009042199

Publication of this book is made possible in part by Santee Cooper.

Halftitle page: Detail of a river craft on the Catawba River from "A Survey of the Establishment of the United States at Mount Dearborn" (1809) by Alexander Macomb. *From American State Papers, Military Affairs*, 3, no. 362. (*Title page*) Detail of John Christian Senf's map of the Santee Canal, as copied by J. P. Gaillard. From U.S. Senate, Ex. Doc. no. 60, *Letter from the Secretary of War* (1881)

To Perry

Contents

Illustrations

Color Plates (*following p. 100*)

Figures

TABLES

Acknowledgments

For help with the research on this book, I am indebted to Carey Lucas Nikonchuk, research consultant at the South Carolina Historical Society in Charleston, and to the other staff of the society, as well as to the staff of the South Carolina Department of Archives and History in Columbia. For access to Dr. Charles Drayton's diary entries on the Santee Canal, I am appreciative of the assistance rendered by Craig Hadley, director of education and research, and Joyce Keegan, collections manager, at Drayton Hall, National Trust for Historic Preservation. I am also grateful to Mary Seabrook Bell, historic-education coordinator at Old Santee Canal Park, for providing me with a copy of Robert B. Bennett Jr. and Katherine H. Richardson's "History of the Santee Canal, 1785–1939"—the basis for establishing Old Santee Canal Park. Mark M. Newell and Joe J. Simmons III of the South Carolina Institute of Archaeology and Anthropology at the University of South Carolina were very gracious in sharing the results of their research on the Santee Canal.

I am indebted to scholars who have gone on before. This publication would not have been possible without David Kohn and Bess Glenn's extraordinary collection of reports, *Internal Improvement in South Carolina, 1817–1828,* privately published in 1938. The documents they gathered into their book were collected by George L. Champion and were later owned by Alexander Young, from whose descendants David Kohn purchased them. Professor F. A. Porcher's paper on the Santee Canal is also extraordinary. First presented in 1875, it was published in 1903 and reprinted several times thereafter. Few papers have an equal published life. Porcher's paper is also important because he would have known the Santee Canal when it was operational. Colonel Arthur Wade's two-part article on Mount Dearborn for the *South Carolina Historical Magazine* is the definitive work on that ill-fated armory. The other excellent scholarship that forms the basis for this publication is included in the bibliography.

I am indebted to Joan Pennington of Penworks for the illustrations superimposing the location of the various canals and waterways onto the extraordinary county maps prepared by Robert Mills (1825). For annotations to the map of the Santee Canal printed in the 1881 U.S. Army report, I am indebted to my stepdaughter, Sarah Prestemon. Kevin Williams of the Tulane Universities Libraries was helpful in tracking down the sketches that Mills executed while in South Carolina. I am also appreciative of the assistance provided by Joyce N. Baker, curatorial assistant at the Gibbes Museum of Art, in making available Charles Fraser's drawing of the Santee Canal and for the assistance of Rebecca McClure, assistant archivist at the Charleston Museum, in making available Andrew Gibbes's sketch of the a cotton box at the Black Oak Lock of the Santee Canal.

General editing was undertaken by my wife, Perry Elizabeth Kapsch, for which I am deeply grateful and to whom this book is dedicated. I am also deeply appreciative to the staff of the University of South Carolina Press. From the very beginning of this project, I have received enthusiastic support from Alexander Moore, acquisitions editor. I am appreciative of his comments, suggestions, and deep knowledge of the history and culture of South Carolina. I was also fortunate to have Karen Rood as the project editor. Finally I am very appreciative to Pat Callahan for the handsome book design.

1 ~ Canals Come to South Carolina

> Few countries enjoy more favorable situations for inland navigation, than South-Carolina; and few, considering the time of her existence, as a settlement, have received equal encouragement in that respect.
>
> John Drayton, *A View of South-Carolina* (1802)

DURING THE FIRST THIRD OF THE 1800S, canal mania created a waterway-building boom that spread across the United States. By 1835 more than thirty-five projects had linked still-water canals with in-river waterways to create approximately twenty-five hundred miles of canal transportation and thousands more miles of improved river navigation. Some canals—such as the Erie Canal, opened in 1825—were eminently successful. Others were not.

South Carolinians were enthusiastic participants in the boom. While New York was building the Erie Canal and the Champlain Canals and Pennsylvania was building the Pennsylvania Main Line Canal, South Carolina undertook a massive program to construct canals and to improve waterways throughout the state. It is said that South Carolina spent more money per capita on internal improvements at this time than any other state. By the end of the 1820s, South Carolina was boasting that it had opened twelve hundred miles of waterways for steamboat navigation and another twelve hundred miles of canals and in-river waterways for small-boat navigation.

As a result of this ambitious program, most South Carolinians had water-transport access to Charleston, the commercial heart of the state and its principal seaport. Steamboats left Granby, near Columbia, for Charleston via the Congaree and Santee rivers. They also left Camden and Stateburg via the Wateree River and Cheraw via the Great Pee Dee River to travel to Charleston. For a few years canal boats left for Charleston from Ware Shoals (then in Abbeville County) via the Saluda River, from Lockhart in Union County via the Broad River, and from above Landsford in Chester County via the Catawba and Wateree rivers. Once well known, South Carolina canals such as Drehr's Canal, Bull Sluice Canal, Rocky Mount

Canal, Saluda Canal, Wateree Canal, and Lockhart's Canal have been are long gone and almost totally forgotten (see color plate 1).

Transporting Cotton

The canal mania was based on sound economy. South Carolina built canals and improved waterways because water transportation was faster, more reliable, and less expensive than road transportation. Most of the existing roads of the time were small, dirt tracks through the wilderness. They were few in number and frequently impassable. The state's chief commodity was cotton, which usually needed to be shipped in bales from the upcountry plantations to Charleston. These bales were unwieldy, heavy, and difficult to move by road. A typical cotton bale weighed five hundred pounds and was almost six-feet long by four- to six-feet wide. Transporting heavy cotton bales on corduroyed roads (built from log planks laid on dirt) and across lowcountry swamps and creeks was expensive, requiring teams of four horses and multiple teamsters. Even for the single traveler unencumbered with cotton, overland travel was difficult. Take, for example, the problems experienced by a man who tried to travel from New York to Charleston by road in 1818:

" Raleigh, (Wednesday) March 18

" Dear Sir—I had flattered myself that I should ere this have had the pleasure of seeing you, for I never could have supposed, for a moment, that I should have met with the delays which I have encountered in my journey here: had I been aware of them, not any consideration of gain, should have tempted me to have undertaken it. I was delayed at Alexandria more than a week, as the roads were impassible I sat out for Richmond and Petersburgh, in the last place I was detained two days for the want of a carriage as the mail only carries four passengers—I offered any money for a private one, but could not procure it for any sum. I have been in this place ever since Monday. Indeed I found it necessary to rest, as my health has lately been very indifferent. It was my intention to have been on the road yesterday, but am prevented by the very heavy fall of snow and the severe cold weather, which from the delicate state of my lungs, since I ruptured a blood vessel, I dare not encounter. I wrote to Mr. Gilfert saying that I should start again, as yesterday. I shall feel much distress if this delay should be injurious to the Theatre, but it is unavoidable. The expences that I have been at, have been great beyond calculation, and I have suffered severely from fatigue and bad roads, in this *dreadful long journey*—besides, my mind has been in a continual state of anxiety since I left New York, which was on the 16th of February. I will thank you to see Mr. Gilfert, and explain to that gentleman the situation I am now in, and assure him, I will not delay one hour unnecessarily, on his account as well as my own. I was in hopes that about this period, I should have finished my time in Charleston, and had arranged my plans accordingly, but, being a stranger, I was ignorant of the time it would take to perform my journey.

" I am, dear Sir, yours truly.

" C. INCLEDON."

Southern Patriot (Charleston, S.C.), March 21, 1818

Road travel was also dangerous. The mail stage frequently issued guns to their passengers because of the danger of highwaymen and other predators along the route.

Water travel could also be treacherous. Boats sank from hitting submerged sawyers (logs stuck in the river bottom and projecting upward). Or boats ran aground on shifting sandbars in rivers. But water travel was seen as the more convenient and safer choice. It was also the most inexpensive means of moving cotton to Charleston. In 1822 the South Carolina Board of Public Works provided an example of the cost efficiency of transporting cotton by water:

> A Broad river boat with four hands, will bring down to Columbia, the distance of 60 miles, 40 bales of cotton in less than two days. The same quantity by waggon, will require at least 4 hands, 16 horses and 4 waggons; and three days to perform the trip: the boat will cost less than $100, and the 4 waggons and 16 horses at least $2,000—the difference of capital employed will be $1,900—the difference of current expense will be the support of 16 horses. If this capital and expenditure were employed for agricultural purposes, it is evident that the farms of the upper country would be vastly improved, and the extent of crops encreased.[1]

South Carolina's program for constructing of canals and improving waterways was an extension of a water culture that already existed. Before the 1820s, even without the canals, South Carolina's rivers were being used to carry cargo during the high waters of spring. Watercraft called Camden boats, Wateree boats, or Congaree-built boats were used for shipping. These boats were large, sixty to seventy feet long and seventeen or eighteen feet wide, with a shallow draft of four to four and a half feet. Such boats could carry 100 to 120 bales of cotton, fifteen to twenty-five cords of wood, or up to sixty hogsheads of tobacco. They sometimes carried sail but were usually poled. In addition rafts and other sorts of boats were found on South Carolina's rivers. Almost always manned by slaves, there riverboats typically required a crew of six to seven African Americans, sometimes more. Experienced crews were valued for their understanding of river navigation and were usually rented or sold as a team.

For all its pluses, South Carolina water transport was localized because it required portages around river obstacles and was reliable only during seasons of high water. With the economic boom following the War of 1812, there developed

SAVANNAH, Feb. 27.

The roads in the upper part of the state are in a dreadful situation, A letter from a correspondent states that between Milledgeville and Augusta, he passed not less than 126 waggons, loaded with cotton, twelve of which were fast in the mud, and two or three upset. The stage between Augusta and Milledgeville was in continual danger of upsetting—*Georgian*.

Southern Patriot, March 1, 1822

We understand that persons who are compelled to travel along the road over the New Bridge, are continually robbed by the gang of armed runaway Negroes, who, as we mentioned a few days past, are hovering about that place Several depredations were committed on Saturday last by these fellows. Eleven of them were seen near the 8 mile post, armed with muskets, cutlasses and hatchets. We trust that prompt measures will be adopted to put a stop to these shameful robberies.

Southern Patriot, November 11, 1822

UNPLEASANT NEWS!!

We are indebted to the politeness of Dr. Miller, (of South-Carolina,) for the following unpleasant intelligence, which he heard at Fort Mitchell a few days ago. In consequence of information received from the commanding officer at Fort Scott, of the scarcity of provisions and dissatisfaction of the troops, General Gaines, accompanied by Major Wright, and several soldiers, embarked in a small boat at Fort Early, with the intention of going to Fort Scott; but when within about 40 miles of their place of destination the boat struck a sawyer, and precipitated all on board in the river. Major Wright and some of the soldiers were drowned. General Gaines reached the western, and one of the soldiers the other bank of the river. The latter repaired to Fort Scott, with the news of this disaster, and Major Bee, with a party of soldiers immediately went in search of the General. They found written with a pencil on a tree, near where the boat was wrecked, his intention of proceeding to Fort Gaines, at which place he had not arrived five days after the boat was wrecked. It is generally feared he has fallen into the hands of the enemy.

[*Milledgeville Geo. Reflector*, 11th inst.

Southern Patriot, March 18, 1818

To be hired the by month or quarter,
SEVEN AS PRIME
Boat Negroes,
As any upon Santee river Enquire of the printers.
August 27 *tw* 6t

City Gazette (Charleston, S.C.), August 31, 1791

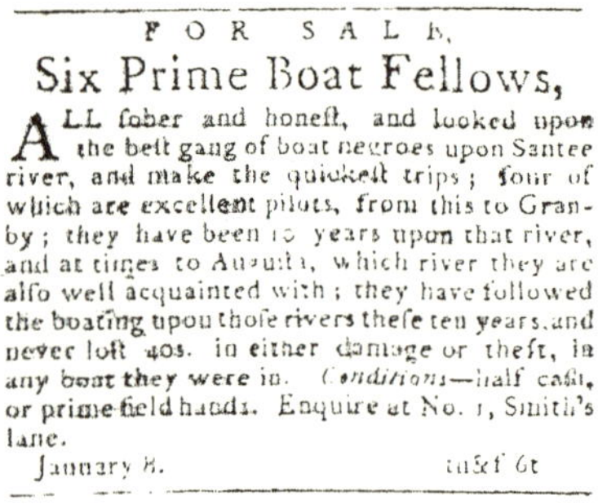

FOR SALE,
Six Prime Boat Fellows,
ALL fober and honeft, and looked upon the beft gang of boat negroes upon Santee river, and make the quickeft trips; four of which are excellent pilots, from this to Granby; they have been 10 years upon that river, and at times to Augufta, which river they are alfo well acquainted with; they have followed the boating upon thofe rivers thefe ten years, and never loft 40s. in either damage or theft, in any boat they were in. *Conditions*—half cafh, or prime field hands. Enquire at No. 1, Smith's lane.
January 8. tu&f 6t

City Gazette, January 11, 1793

an urgency in South Carolina to undertake a canal-construction and waterway-improvement program. The price for cotton had increased, but Charleston's share of the cotton-export market seemed to be shrinking when compared to that of its rival, Savannah. Savannah enjoyed a shipping advantage over Charleston because planters along the Savannah River could easily ship cotton down the river to Savannah for export. Although Charleston is situated on two tidal rivers, the Ashley and the Cooper, they do not penetrate far into the interior of South Carolina. Charleston, unlike Savannah, did not have a natural water course allowing it easy access to upcountry cotton. The leading citizens of South Carolina decided that something needed to be done. There were three means of improving the shipment of cotton from the upcountry to Charleston: more passable roads, a still-water canal, and a river/canal navigation system using South Carolina's rivers. Railroads, barely in their infancy in England, were not yet a viable alternative.

Roads did not seem a reasonable option. The few rudimentary roads of South Carolina were usually corduroyed, built with local pine trees that rapidly decayed in the warm, moist climate of South Carolina's lowcountry. Bridges were rare, usually made of wood that also decayed rapidly in this era before chemical preservatives. Where there were bridges, they were frequently knocked out by floods.[2] The levy imposed on slave owners required slaves to be provided to state authorities for the construction and maintenance of roads and other public facilities. Roads constructed by the levy were rarely well built.[3] To improve existing roads would have required stone to replace the decaying corduroy. At that time South Carolinians believed that there was no source of stone in the lowcountry. Transporting stone from the upcountry would be prohibitively expensive. Improving the existing roads from the upcountry all the way to Charleston would also have required the construction of expensive causeways and large-span bridges to cross the lowcountry's numerous boggy swamps and tidal streams (see fig. 1.1).

The second option was a still-water canal, an excavated ditch filled with water from specially constructed reservoirs or from adjacent waterways dammed to divert water into the canal. Changes in elevation of the canal were controlled by a series of locks. Although expensive to build, a still-water canal provided the most reliable means of water transport. At the time when South Carolina was considering a statewide canal-construction and waterway-improvement program, architect-engineer Robert Mills[4] proposed a still-water canal from Columbia to Charleston.[5] But the expense of such a still-water canal was an issue. Excavation of a canal from

Fig. 1.1. Transporting cotton on a corduroyed road. Illustration by Edwin Forbes for *Harper's Weekly* (May 12, 1866). Courtesy of the Library of Congress

Fig. 1.2. Languedoc Canal in France opened in 1681 and was the prototype for still-water canals throughout eighteenth- and nineteenth-century Europe and America. First published in Denis Diderot's 1751 *Encyclopédie,* this illustration shows the operation of two lift locks, with lock gates, balance beams, and gate valves for raising (or lowering) boats.

Columbia to Charleston, equipped with expensive masonry locks, diversion dams, and other appurtenances would require large expenditures of funds (see fig. 1.2).

A navigation system that combined canals with South Carolina's extensive rivers was less expensive than Mills's proposed still-water canal and therefore very attractive to South Carolinians. The central portion of South Carolina was drained by the large rivers of the Saluda-Broad-Congaree-Wateree-Santee river system. Immediately above Charleston was the Santee River. Since 1800, boats from the Santee River

Fig. 1.3. An 1881 depiction of Charleston in 1780. From William Cullen Bryant and Sydney Howard Gay, *A Popular History of the United States* (1881–83)

had been reaching Charleston via the Santee Canal and the Cooper River. If cleared of obstructions, the Santee, Wateree, and Catawba rivers could provide access to the center portion of the state. Further west, the Santee River also gave access to the Congaree River, which in turn gave access to the Broad and Saluda rivers, which drained the central and western portions of the state. Obstructions in all these rivers, it was thought, could be cleared at relatively little cost. Small bypass canals could take boats around obstructions such as rapids and falls.

Clearing the Saluda-Broad-Congaree-Catawba-Wateree-Santee river system of obstacles would provide downriver access to the upper end of the Santee Canal and through that canal to the Cooper River and Charleston. The Santee Canal became the critical link in South Carolina's canal-construction and waterway-improvement program, one the most extensive of such programs yet undertaken in the United States.

The Canal Boom

South Carolina's ambitious program to build canals and improve waterways throughout the state was part of the canal boom that was sweeping America. There were two periods of extensive canal and waterway construction. The first canal construction period began at the end of the Revolutionary War, in 1783, and lasted until the beginning of the War of 1812, during which large construction projects were not undertaken. The second took place after the end of the War of 1812 in 1814 and lasted until the outbreak of the Civil War in 1861. Most of South Carolina's canals and water navigations were built during the second period.

During the first construction period, canals built in the United States tended to be short and were not financially successful. The first canal constructed in Pennsylvania, for example, was the Conewago Canal (built 1792–97), which was slightly under a mile long. In Maryland the longest canal from before the War of 1812 was the Potomac Canal, built in 1785–1802 by the Potomac Canal Company. It extended 184 miles from Georgetown to Cumberland, Maryland. Through most of its course, it was a river navigation where channels in the riverbed were improved by in-river sluices and other devices. The only true canals built by the Potomac

Canal Company, those that contained lift locks, were at Great Falls (.68 miles long) and at Little Falls (2.16 miles long). The financial success of Potomac Canal was uneven. In Virginia the longest canal from before the War of 1812 was the James River Canal. Like the Potomac River Canal, the James River Canal was an improved river-navigation system 220-miles long with still-water canals only at the upper falls of the James at Richmond (completed in 1789) and at the lower falls (completed in 1796). These were consolidated in 1810 with a canal around Richmond. Early canals that were excavated still-water canals and did not use the riverbed for navigation tended to be short. The Middlesex Canal in Massachusetts (built 1789–1809) was only 30 miles long. In South Carolina, the Santee Canal (built 1793–1800) was 22 miles long.

The canals constructed during the second canal period, after the end of the War of 1812, were much longer and more often were excavated canals (also called still-water canals) equipped with locks. The Erie Canal (built 1817–25) was 363 miles long; the Ohio and Erie Canal (built 1825–32) was 307 miles long; the Chesapeake and Ohio Canal (built 1828–50), the successor to the Potomac Canal, was 184 miles long (and was planned to be twice that length); and the James and Kanawha Canal (built 1830–51), the successor to the James River Canal, was 196 miles long.

During the first period canals were typically built by private organizations chartered or otherwise authorized by the state. In 1792 Governor Thomas Mifflin of Pennsylvania authorized Robert Morris and his group to build the Conewago Canal. The Potomac Canal Company, a private company, was authorized by the Virginia (1784) and Maryland (1785) legislatures. The James River Canal Company, another private company, was authorized by the Virginia legislature (1784). On March 22, 1786, the South Carolina legislature passed "An Act to Establish a Company for the Inland Navigation from Santee to Cooper River." The official name of the company was "The Company for the Inland Navigation from Santee to Cooper River," more popularly known as the Santee Canal Company.

During this first period states sometimes provided small amounts of money to help build canals. The State of Pennsylvania provided £5,250 (or $14,000) to Robert Morris and his group to build the Conewago Canal—a small fraction of the officially estimated total construction cost of $119,000. Usually funding for early canals came from the shares purchased by proprietors or stockholders, sometimes augmented with whatever could be raised by lotteries. In capital-starved early America canal construction was frequently halted by a lack of money, as happened with the Chesapeake and Delaware Canal in Delaware and Maryland and the Susquehanna Canal in Maryland. The early canals of South Carolina, such as the Santee Canal, were also state chartered and privately funded. And, as elsewhere in America, lack of money caused some canals in South Carolina—such as those of the Catawba and Wateree Navigation Company, the Edisto Company, and the Sampit and St. James Santee Canal Company—not to be completed.

After the War of 1812, canals were frequently financed and constructed by the individual states, as occurred in New York (the Erie Canal and the Champlain Canal), Pennsylvania (the Pennsylvania Main Line Canal and its divisions), Virginia (the James and Kanawha Canal) and in South Carolina. During this second period, even when canals were privately funded, states frequently bought shares of the canal company, as happened with the Chesapeake and Ohio Canal Company.

The history of South Carolina's canals and waterways begins with its first completed canal, the Santee Canal, chartered in 1786 and opened in 1800. This canal's success was thwarted because the Santee, Congaree, and Wateree rivers above the entrance to the Santee Canal were frequently blocked with obstructions and sandbars, thereby limiting the number of boats that could descend these rivers and use the Santee Canal. South Carolina's decade-long program (1818–28) to build canals and clear its rivers of obstacles to navigation resulted in large increases in shipping on the Santee Canal. These new canals and navigable waterways also decreased shipping costs while increasing both the volume of agricultural produce (especially cotton) shipped and the wealth of the state. But these canals and waterways tended to be short lived. By the end of the 1820s, a new transportation system based on the railroad and turnpike had begun to be developed in South Carolina and elsewhere in the United States. This system replaced and led to the decline of canals and waterways throughout the country.

Transporting Cotton from the Upcountry

The impetus for building canals and improving rivers for navigation originated with the geography of South Carolina. Like many of the original thirteen states, South Carolina is a coastal state located between the Appalachian Mountains to the west and the Atlantic Ocean to the east. Initial settlement was along the coast. As long as South Carolina settlement was limited to a narrow coastal strip of lowcountry approximately one hundred miles wide, transportation was easy. Tidewater streams and rivers reached most lowcountry plantations, each of which maintained its own dock and communicated with other plantations and coastal cities by water. Plantations had their own boats and boat crews, typically with four to six oarsmen. In addition shipping along the South Carolina coast, the coasting trade, was reliable and well established.

The focal point of South Carolina's coastal region was the port city of Charleston. Produce grown in the lowcountry was shipped to Charleston. Situated on two tidal rivers, the Ashley and the Cooper, Charleston had an excellent harbor. Its location in the south of the United States made it well suited for trade with the Caribbean, as well as with Europe and with the other Atlantic coast colonies.

After the Revolutionary War, the economy of South Carolina started to change radically with expansion of cotton production, which began in the 1790s and grew rapidly after the introduction of the cotton gin in 1802 (see figs. 1.4 and 1.5). Most of the new cotton production took place in the upcountry of South Carolina with its inexpensive lands and optimal growing conditions. Upcountry cotton plantations, however, were usually one hundred miles or more from the coast, creating the urgent need to develop economical means of transporting cotton to Charleston for shipment overseas and, to a lesser extent, to northern American ports.

After the Revolutionary War exports of cotton exploded. Exports of indigo withered, and those of rice remained stable. This can be seen in the statistics complied by John Drayton for *A View of South-Carolina, As Respects Her Natural and Civil Concerns*, published in 1802 (see tables 1.1 and 1.2).

The explosive growth of upcountry cotton after the Revolutionary War was paralleled by the decline of one of South Carolina's traditional cash crops, indigo, and the transformation of the other, rice. After the war the British bounty (bonus) on indigo was eliminated. Competition increased from the West Indies and India. By 1800 almost no indigo was being exported from South Carolina. Former indigo planters in the lowcountry turned to rice production; those in the upcountry turned to cotton.

TABLE 1.1. "Statement of exports from the Port of Charleston, and their total annual amount, from 1790 to 1800"

Period of Export.	*Barrels of Rice.*	*Casks of Indigo & lbs weight.*	*Hbds. of Tobacco.*	*Pounds wt of cotton.*	*Total annual of exports, including Rice, Indigo, Tobacco & Cotton.*
					Dollars.
From Aug. 31, to Dec. 31, 1789, being 4 months.	9,157	289	1,172	5,670	313,674
1790	87,179	1,649	6,820	9,840	2,104,677
1791	98,044	2,245	6,033	54,075	2,411,771
1792	102,235	2,495	5,285	76,710	2,341,777
1793	94,055	1,819	3,324	93,540	3,112,525
1794	69,717	2,154	4,908	159,040	3,869,019
1795	85,670	1,217	4,288	1,109,653	5,984,196
1796	84,540	490	5,328	912,600	7,600,387
1797	80,837	96,121 lbs.	3,961	1,008,511	6,459,524
1798	74,277	19,838 lbs.	4,638	2,476,431	6,946,924
1799	70,426	6,892 lbs.	9,640	2,801,936	8,729,015
1800	75,788	3,400 lbs.	7,927	6,425,862	10,554,802

The years comprise periods from October 1 in one year to September 30 in the next.

Source: John Drayton, *A View of South-Carolina* (1802), 168.

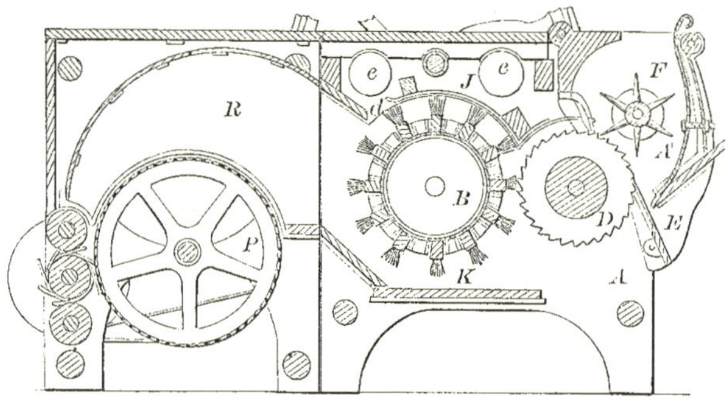

Cotton-Gin.

Fig. 1.4. Section drawing of a cotton gin. From Edward H. Knight, *Knight's Mechanical Dictionary* (1877). Placed in the hopper (F), cotton is subjected to the rotating action of the picker roll (A). The rotating saw blades (D) force the cotton through a grid (E). This action separates the fiber lint from the remainder of the plant. The rotating saw blades are brushed with rotating brushes (B). The heavier debris falls to the board labeled (K). The lighter lint and dust are passed to chamber on the left (R). The dust passes through the wire-mesh surface of a rotating cylinder (P) and out of the machine. The lint is passed on to the doffing and compacting cylinders.

Fig. 1.5. "Scenes of a Cotton Plantation" (*right and facing*), by Alfred Rudolph Waud (*Harper's Weekly*, February 2, 1867). Courtesy of the Library of Congress

Rice production in the postwar years was greatly changed. Tidal cultivation of rice replaced the traditional lowcountry swamp plantations. Tidal cultivation could produce five to six times the slave yield compared to the inland swamp plantations. During the early decades of the nineteenth century, many of the traditional rice plantations of the lowcountry became abandoned.[6] In 1825 Abram Blanding,[7] superintendent of public works for South Carolina, reported on the decline of rice production and the depopulation of the lowcountry of the Edisto River valley south of Charleston:

The inland rice swamps of *Stono* and *Cacaw* thirty years ago, were generally under cultivation. There were one hundred and nineteen settled plantations there, containing together twenty-four thousand acres of highly improved rice lands, usually producing two barrels or one thousand four hundred pounds of rice to the acre, of a quality superior even to that made in the tide swamps. At present there are about

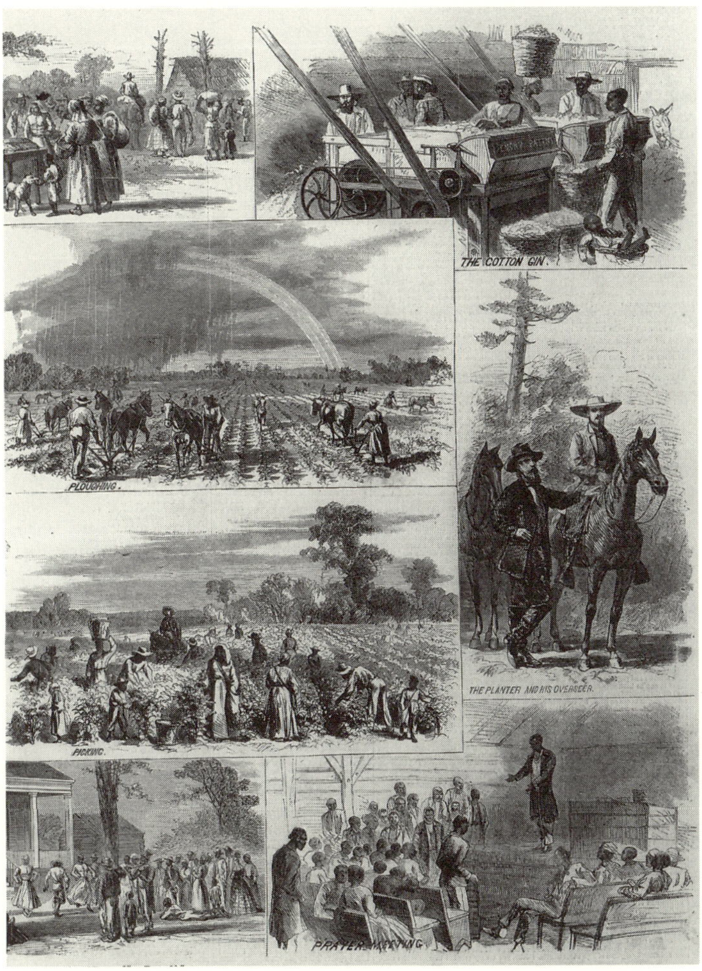

TABLE 1.2. "A Statement, shewing the quantities of Rice, Indigo, Tobacco, and Cotton Exported from South-Carolina at different periods, from the year 1720, to the year 1801, inclusive"

Years.	*Bbls. of Rice.*	*lbs. weight of Indigo.*	*Hhds. Tobac.*	*lbs. weight of Cotton.*	*To what place exported.*	*Value of Exports*
1720 to 1729*	263,488				Foreign parts	
1730 to 1739	499,525				Ditto.	
1731	41,957				Ditto.	
1733	36,584				Ditto.	
1739	71,484				Ditto.	
1740	91,110				Ditto.	
1747		200,000			Great Britain alone.	
1754	104,682	216,924		Some Cotton	Foreign parts.	£.241,529 00s 00d.
1760 to 1761	100,000	399,366	11		Ditto.	£.256,767
1761 to 1762	62,288	249,000			Ditto.	
1767 to 1768					Great Britain alone.	£.508,108 6s. 10d.
1768 to 1769					Ditto.	£.387,114 12s. 1d.
1769 to 1770					Ditto.	£.278,907 14s.
1770 to 1771					Ditto.	£.420,311 14s. 8d.
1771					Foreign parts.	£.756,000 1s. 1d.
					Great Britain	£.456,513 8s. 4d.
1772	140,000				Foreign parts.	
		1,107,660			Great Britain alone.	
1782	24,224	827 cask	613		Foreign parts.	
1783	61,974	2,051 do.	2,680		Ditto.	
1789	100,000				Ditto.	
1792	106,419	839,666 lbs.	5,290	68,520	Ditto.	$2,917,979
1795	85,670	1,217 casks	4,288	1,109,653	Ditto.	$5,984,196
1799	70,426	6,892 lbs.	9,646	2,801,996	Ditto.	$8,720,015
1800	75,788	3,400 lbs.	7,927	6,425,863	Ditto.	$10,554,842
1801	64,769	8,502 lbs.	5,996	8,201,007	Ditto.	$14,304,045

*In the year 1724, 18,000 barrels of rice, 52,000 barrels of pitch, tar, and turpentine, together with deer skins, furs, and raw silk, were exported to England. Besides which, rice and other articles were exported to the West Indies, and the northern states of America.

Source: John Drayton, *A View of South-Carolina* (1802), 173.

twelve of these plantations cultivated only in part; the rest are now entirely abandoned. Their banks, the work of more than a century, are broken and washing down, and their canals and drains are filling up. The great public drains which lead through these swamps, opened formerly by the combined labor of these plantations, are now choaking up, and are already unfit for the purposes for which they were constructed.[8]

Blanding's report provides a melancholy description of the effect of depopulation: "Ruins are every where to be seen. Splendid buildings once the residence of industry, wealth and unbounded hospitality, are now deserted and crumbling to the dust, or have already reached that period of decay, which has left nothing but a few mouldering ruins to mark their former site."[9]

Blanding provided five reasons for the depopulation of the lowcountry:

1st. The abolition of the rights of primogeniture.[10]

2nd. The high price and consequently profitable culture of cotton, when it was first introduced.

3rd. The uncertainty of the rice crop in consequence of the failure of the reservoirs in dry years, and the imperfect drainage which in wet years subjected the crops to floods.

4th. The deterioration of the soil from long unskillful and uninterrupted cultivation.

Lastly—The unhealthiness of the country occasioned by the stagnant water and vegetable decomposition of abandoned rice fields.[11]

There were two types of cotton grown in South Carolina: Sea Island (also called black-seed or long-staple) cotton and short-staple (also called upland or green-seed) cotton. Sea Island cotton grew well in the coastal areas, and it was relatively easy to separate seeds from the fiber by hand. In contrast short-staple cotton proved difficult to separate lint from the seed. Only with Eli Whitney's invention of the cotton gin in 1793 (see fig. 1.4) did cotton production become economically profitable. And cotton proved very lucrative. Lowcountry plantation owners converted to cotton production and developed upcountry cotton plantations. Upcountry farmers increasingly grew cotton as a cash crop and became wealthy as a result. Inexpensive land prices in the upcountry were also a major factor in the expansion of cotton production in that region (see table 1.3). South Carolina entered its first cotton boom (1794–1819).

As cotton replaced rice as South Carolina's principal cash crop, the geographic center of agricultural production of the state moved westward. This changed South Carolina's transportation needs. When rice was the predominant crop, the lowcountry's rice plantations, particularly the tidal plantations, were close to navigable streams and water transport to Charleston was readily available. With cotton increasingly being grown in the central and western portions of the state, easy water navigation to transport agricultural produce to Charleston was no longer

TABLE 1.3. Land costs in South Carolina, 1802

Land	*Cost*
"Best lands in this state, which are tide swamps, if cultivated, have sold for one hundred and seventy dollars an acre."	$170 per acre
"In general, however, they sell from seventy to ninety dollars an acre; on a credit of one or two years."	$70 to $90 per acre
"Uncultivated tide lands sells proportionately lower."	
"Inland swamps, if cultivated, sell at prices betwixt twenty and fifty dollars each acre."	$20 to $50 per acre
"Good cotton land, has sold in Beaufort district, as high as sixty dollars per acre."	$60 per acre
"In general, however, its value, in different parts of the state, is from six, to forty dollars; the same depending much on its situation; as that nearest the sea, is considered the most valuable, and produces the finest cotton."	$6 to $40 per acre
"Other high lands, sell from one to six dollars an acre; according to their respective situations, and conveniences to navigation. Hence, men possessing any capital whatever, may settle themselves indepen-dently; upon lands which descend to their posterity; together with every improvement made thereon; by their industrious labour."	$1 to $6 per acre

Source: John Drayton, *A View of South-Carolina* (1802), 110–11.

possible. The falls and rapids along rivers of the fall line, which separated the lowcountry from the upcountry, hindered water travel. Below the fall line, rivers such as the Congaree, the Santee, and the Great Pee Dee were used for shipping but only during the high waters of spring and only with difficulty because of sandbars, logs, and other obstructions. Road access to Charleston from the upcountry was also difficult. The few roads that did exist had to traverse the swamps and the broad tidal rivers and creeks of the lowcountry and were usually not well constructed or well maintained.

The Political Base for Internal Improvements

South Carolina's massive and expensive initiative to improve transportation from the upcountry to Charleston required broad political support within the state. The political base for the initiative was provided by merchants, planters, farmers, and artisans—mostly white males. By the 1790s—by the time South Carolina's economy began to turn from rice toward cotton and from the lowcountry toward the upcountry—the social structure of the state had been established.

At the top were the merchants, commodity brokers, and planters of Charleston. Drayton reported in 1802 that some of these lowcountry planters received an annual income from their plantations of eighty thousand dollars per year, a princely sum in the 1790s. Most received twelve thousand to twenty thousand dollars a year, and the greatest number of planters received at least six thousand dollars per year—in a period when a skilled workman was typically paid a dollar a day. The planters had disposable income to indulge in luxuries unavailable to the rest of the population. As Drayton reported, lowcountry planters sent to Europe for gardeners to take care of their South Carolina estates, the value of which varied from twenty thousand to forty thousand dollars.

At this time lowcountry planters were the principal owners of African American slaves. In 1802 Drayton provided a contemporary South Carolinian's view of the need for the institution of slavery in South Carolina:

> In the pursuits of agriculture, slaves were introduced into this state; and importations from Africa soon supplied the planter with as many negroes, as he was able to purchase. This gave a rapid encrease to the settlement and riches of the lower country; when, otherwise, its richest lands, would not have been worth the cultivating. They, consequently, became a vested property in their respective owners, by the laws of the land; and however paradoxical it may appear, their owners, on obtaining their independence, and a right by the constitution and government of this state, and these United States, thence flowing, *to be protected in their person and property,* had an indefeasible right in them; without the reach of law to alter, unless by their own consent, or by suitable compensation.[12]

As cotton replaced rice as the state's principal cash crop, many rice planters diversified to cotton. Some relocated from the lowcountry to the upcountry, but others who established upcountry plantations continued to live in the lowcountry. These planters, with support from Charleston merchants, dominated the state legislature at the time. These same individuals would greatly benefit from an improved system for transporting cotton to Charleston.

The next economic class was that of the independent working farmer. Any massive canal construction and waterway improvement program within South Carolina would require at least the acquiescence of the farmers of the state. Although not nearly as wealthy as the planter, the independent farmer was well-to-do. Drayton reported that these independent farmers had a yearly income of two thousand

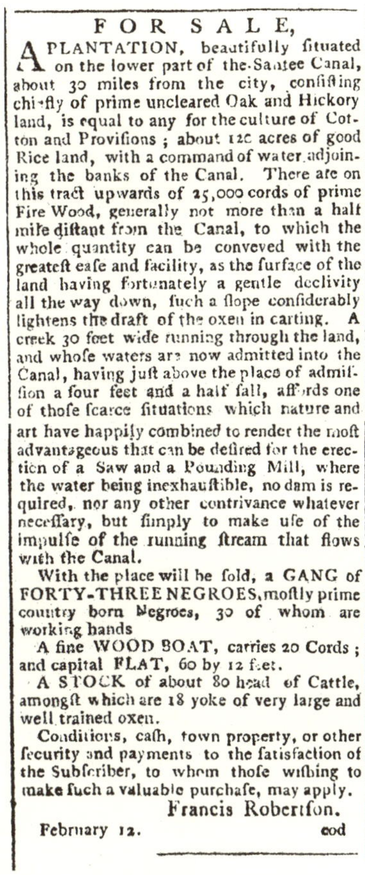

FOR SALE,

A PLANTATION, beautifully ſituated on the lower part of the Santee Canal, about 30 miles from the city, conſiſting chiefly of prime uncleared Oak and Hickory land, is equal to any for the culture of Cotton and Proviſions; about 120 acres of good Rice land, with a command of water adjoining the banks of the Canal. There are on this tract upwards of 25,000 cords of prime Fire Wood, generally not more than a half mile diſtant from the Canal, to which the whole quantity can be conveved with the greateſt eaſe and facility, as the ſurface of the land having fortunately a gentle declivity all the way down, ſuch a ſlope conſiderably lightens the draft of the oxen in carting. A creek 30 feet wide running through the land, and whoſe waters are now admitted into the Canal, having juſt above the place of admiſſion a four feet and a half fall, affords one of thoſe ſcarce ſituations which nature and art have happily combined to render the moſt advantageous that can be deſired for the erection of a Saw and a Pounding Mill, where the water being inexhauſtible, no dam is required, nor any other contrivance whatever neceſſary, but ſimply to make uſe of the impulſe of the running ſtream that flows with the Canal.

With the place will be ſold, a GANG of FORTY-THREE NEGROES, moſtly prime country born Negroes, 30 of whom are working hands

A fine WOOD BOAT, carries 20 Cords; and capital FLAT, 60 by 12 feet.

A STOCK of about 80 head of Cattle, amongſt which are 18 yoke of very large and well trained oxen.

Conditions, caſh, town property, or other ſecurity and payments to the ſatisfaction of the Subſcriber, to whom thoſe wiſhing to make ſuch a valuable purchaſe, may apply.

Francis Robertſon.

February 12. eod

City Gazette, February 12, 1800

dollars upward. Although one might own one or two slaves, independent farmers were generally not slave owners. Clearly the yeoman farmer was the favorite of Drayton's: "But, it is here, that health, and independence dwell. And a crop of an hogshead of tobacco, or a bag or two of cotton, forms an income, which pays the taxes and expences of the farm; and makes a family happy and contented."[13]

Farmers with large land holdings and excess crops to sell, would particularly benefit from an improved transportation system. In addition to cotton, wheat was grown in the upcountry and shipped to Charleston for export overseas, especially the Caribbean. When the price of cotton fell precipitously in the early 1820s, many planters and farmers switched from cotton to wheat. Like planters and merchants, these independent farmers supported South Carolina's internal-improvement program.

Farmers with small land holdings were less dependent on improved transportation to Charleston, but the Board of Public Works developed a justification of why these farmers should also support the canal-construction and waterway-improvement program, pointing out that these farmers needed salt for the preservation of their meats. This salt was imported to Charleston from Turks Island and elsewhere in the Caribbean and then shipped to the upcountry. According to the board, the decrease in the shipping costs of salt alone that would result from improved water transportation would save the farmer enough money to cover the taxes he owed on his a five-hundred-acre farm (thought to be a typical small farm), which were seventy-five cents per year.[14] Small farmers could also expect to pay reduced prices on manufactured goods shipped to the upcountry by improved, less expensive transportation. These arguments helped to solidify farmers' political support of South Carolina's canal-construction and waterway-improvement program through most of the 1820s.

South Carolina also had a class of citizens that might be called artisans, workers who went through formal apprenticeships of five to seven years and entered such trades as stonemason, bricklayer, and carpenter. Typically these workers resided in the cities and towns of South Carolina. Although an artisan's annual income was much less than either a planter or a farmer, he earned enough to support a family and own a modest amount of property. Because South Carolina artisans were few in number, their support was not politically important, but it was strong because the South Carolina canal-construction and waterway-improvement program offered a steady source of well-paying employment. As it turned out, with the scarcity of skilled artisans in the state, many had to be imported from the north to build the canals and clear the waterways of South Carolina.

Within South Carolina there were also "unskilled" laborers, both white and free black. Not having served apprenticeships, they had no trade. They were typically paid two shillings four pence (about fifty cents) per day plus provisions for work on the canal construction and waterway improvements of South Carolina.

They worked six days a week from sunrise to sunset with a two-hour break for the midday meal. Frequently they were unemployed during the winter months. The improvements program provided these laborers with work, but even with steady employment, life was hard for the laboring class (see tables 1.4 and 1.5).

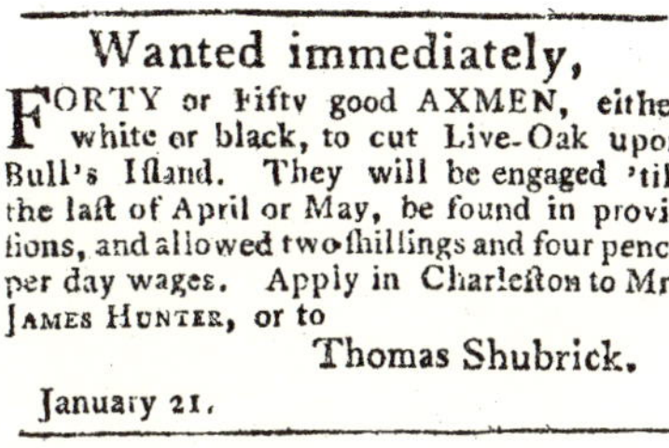

Wanted immediately,

FORTY or Fifty good AXMEN, either white or black, to cut Live-Oak upon Bull's Iſland. They will be engaged 'till the laſt of April or May, be found in proviſions, and allowed two ſhillings and four pence per day wages. Apply in Charleſton to Mr. JAMES HUNTER, or to

Thomas Shubrick.

January 21.

City Gazette, February 4, 1800

By 1802 the population of South Carolina was growing quickly, and—despite a ban on the importation of slaves that had been in effect since the Revolution—a large percentage of that population was black slaves. At the lowest rung of the social and economic order in South Carolina, slaves undertook the bulk of the physical work in building canals and clearing rivers.

Improving the River System

Once it was decided that water transport was the least expensive means for transporting agricultural produce to the coast,[15] South Carolina prepared plans to improve the three major river systems that drained the state: the Great Pee Dee and its tributaries in the north, the Saluda-Broad-Congaree-Catawba-Wateree-Santee river system in the middle of the state, and the Savannah in the south. The Saluda-Broad-Congaree-Catawba-Wateree-Santee river system was the most important for

TABLE 1.4. Economic/social classes of South Carolina, 1802

	Maximum Income	*Minimum Income*	*Value of Estate*
Planters	$80,000 per year. Many receive $12,000 to $20,000 per year. Greatest part receive $6,000 per year.	$3,000 per year	$20,000 to $40,000
Farmers	Income ranges from $40 to $2,000 per year	"But, it is here, that health, and independence dwell. And a crop of a hogshead of tobacco, or a bag or two of cotton, forms an income, which pays the taxes and expences of the farm; and makes a family happy and contented."	
Artisans (masons, bricklayers, shipwrights); overseers; etc.	$1 per day (not addressed by Drayton)	$300 per year (not addressed by Drayton)	
Laborers	$60 per year (not addressed by Drayton)		

Source: John Drayton, *A View of South-Carolina* (1802), 110, 112–13.

TABLE 1.5. Population of South Carolina, 1700–1800

Years	*White*	*Blacks and Persons of Color*	*Total*
1670	A Colony sent over under Gov. Sayle		
1700	5,500		
1721	14,000		
1723	14,000	18,000	32,000
1734	7,333	22,000	29,333
1765	40,000	90,000	130,000
1792	140,178	108,895	249,073
1800	196,255	149,336	345,591

Source: John Drayton, *A View of South-Carolina* (1802), 103.

conveying commerce to Charleston. Since 1800 the Santee River had been navigable to Charleston via the Santee Canal and the Cooper River. As plans for statewide improvements were formulated, it was determined that the Santee Canal would continue to be the critical link in what would prove to be one of the most extensive canal and river-navigation program yet undertaken in the United States.

On a map the Saluda-Broad-Congaree-Catawba-Wateree-Santee river system looks like a misshaped trident, or three-pronged fork. The Saluda River is the western prong, bending westward and joining the Broad River, just above Columbia to form the Congaree River. The Broad and Congaree form the middle prong. The eastern prong starts in the north with the Catawba River, which becomes the Wateree, which joins the Congaree southeast of Columbia to form the Santee River, the handle of the fork. It flows southeast and splits into the North Santee and the South Santee about twenty miles before reaching the Atlantic Ocean, approximately sixty miles north of Charleston.

There were two problems in using the Saluda-Broad-Congaree-Catawba-Wateree-Santee river system for commerce. Narrow, shallow-draft river vessels well suited for the upper reaches of these rivers were ill suited for the sixty-mile Atlantic Ocean passage from either mouth of the Santee River south to Charleston. Such an ocean voyage in these boats could result in the loss of cargo and crew. Moreover the upper reaches of these rivers had extensive falls and rapids that prevented river access downstream at most times of the year.

The answer to both of these problems was bypass canals around the falls and rapids of the Saluda, Broad, Wateree, and Catawba rivers and the clearing of sandbars, snares, and other obstacles to allow the use of larger vessels on these rivers and on the Congaree and Santee during most months of the year. River clearing below the fall line would not only facilitate canal-boat travel but would also permit steamboats to reach the middle of the state from Charleston. This construction and clearing work was undertaken by the state government on a comprehensive basis throughout the state.

From 1818 to 1828 the State of South Carolina undertook a massive program to improve its river systems by clearing obstacles in the great rivers of the South Carolina upcountry—the Saluda, the Broad, the Congaree, the Catawba, the Great Pee Dee, and the Wateree—and by building bypass canals around rapids and falls so that downstream commerce on these rivers could flow through the Santee Canal to Charleston. In addition the state undertook other transportation projects, notably improving coastal navigation and building the State Road from Charleston through Columbia to the Saluda Mountains northwest of Columbia near the North Carolina border.

Organization of the State Internal-Improvement Program

The ambitious program to improve water transportation was directed by John Wilson, the state civil and military engineer in 1818–20.[16] In 1820 the Board of Public Works was founded to supervise the work. Joel R. Poinsett[17] was named board president; Abram Blanding was named acting commissioner of roads, rivers and canals; and Robert Mills, Nicholas Herbemont,[18] and Robert G. Mills[19] were named commissioners. In 1822 the board was superseded by the position of superintendent of public works. Abram Blanding was appointed to that office and occupied it for the duration of the internal improvement program.[20]

Impact of the Program

Initially budgeted at one million dollars, South Carolina's ambitious internal-improvement program ultimately cost almost two million dollars. Some historians have characterized this program of canal construction and improved river navigation as a failure because most of the improvements were abandoned within a decade of their completion. Although this was true, during the years in which the program operated, these improvements greatly increased shipments of cotton and other commodities to Charleston for transshipment overseas. For example the quantity of upland cotton shipped to Charleston doubled between 1820 and 1826.[21]

There were also some lasting benefits from this program. Although most of the canals constructed and river navigations developed by the state were later abandoned, portions of the coastal-navigation improvements remain in existence today and have been incorporated into the Intracoastal Waterway. The third component of the program, the State Road, which some historians have characterized as a failure, was in fact a success and led the way for new roads throughout the state.

South Carolina's efforts in 1818–28 to build canals and open navigable waterways was a massive effort for its time. A new canal technology was being used to develop a transportation system, and the men who formulated and implemented this program had little or no experience in canal construction. South Carolina's canal-building and waterway-improvement program marks the beginnings of the Industrial Revolution in America, These canals and waterways ushered in a new transportation revolution based on the steamboat, railroad, and road. The story of the development and demise of these canals and river navigations helps us to understand the transportation revolution that was taking place throughout the United States.

2~ Building the Santee Canal, 1786–1800

The Santee Canal is said to be at least equal to any work of the kind, in these United States.

John Drayton, *A View of South-Carolina* (1802)

The Santee Canal . . . is of very great importance to the trade and intercourse between the port of Charleston and a large section of the state, as also a part of North Carolina.

George Armroyd, *A Connected View of the Whole Internal Navigation of the United States* (1830)

Santee Canal, connects the harbour of Charleston with the Santee. It commences on the west branch of Cooper river, and passing along Biggin Swamp, intersects Santee river at Black Oak Island. Length, 22 miles; course, N.N.W.; 32 feet wide at top, 20 at bottom; 4 feet deep; rise and fall, 103; 13 locks, each 60 by 10 feet; completed in 1802 [*sic*]; cost $700,000. By means of this canal, and the Santee and Congaree rivers, which have been improved, a navigable communication is afforded from Charleston to Columbia.

Henry S. Tanner, *A Description of the Canals and Rail Roads of the United States* (1840)

The Santee Canal . . . is said to have proved disastrous to those engaged in it.

S. Augustus Mitchell, *Mitchell's Compendium of the Internal Improvements of the United States* (1835)

ALTHOUGH IT IS A VERY EARLY CANAL, the Santee Canal, completed in 1800, is not South Carolina's first. Earlier, smaller canals were undertaken to meet local needs. The General Assembly passed the first South Carolina canal legislation in 1714 to allow construction of a local water course on the lands of John Jones on Edisto Island, southwest of Charleston.[1] This canal was to be built by a levy of all male persons from sixteen to sixty years of age living on the southeast part of Edisto Island. In 1717 an act was passed to "make one or more Cuts or Creeks in the Parish of Christ Church, for the more convenient transportation of the goods carried by water to Charleston."[2] There were also earlier government-sanctioned efforts to clear waterways. In 1726 the legislature authorized the clearing of Biggin Creek from Stoney Landing to Herman's Bridge for navigation purposes.[3] Similar legislation was passed for several other rivers throughout South Carolina.

The origins of the Santee Canal date to as early as 1770. In that year James Cook, reportedly a surveyor, wrote a letter to the publisher of the *South Carolina Gazette,* a Mr. Timothy, proposing a canal between the Santee and Cooper rivers: "The opening a communication from *Santee* to *Cooper* river, by a navigable cut, is very practicable to be performed, and will be productive of infinite advantage to the community."[4] Cook saw the advantages of such a canal for carrying agricultural products from the upcountry to the port of Charleston and for providing water-power sites:

> The interior parts of this province, not being fully known to many of the residents near Charles-Town, it will therefore be necessary to advance a moderate calculation of the additional trade, that might accrue on this occasion. I have reckoned an hundred and twelve millseats on the navigable waters of *Santee* river, above the cut, all situated near to the best of timber and rich lands, And supposing two saws to belong to each mill, sawing four hundred each feet per day, the total amount of the whole, for one day, will be eighty-nine thousand six hundred feet; most of which would be sent from thence to Charles-Town; besides immense quantities of indico [*sic*], hemp, flax, tobacco, madder etc.[5]

He believed the canal would to lead to prosperity: "We should see the waters of those rivers crowded with rafts and boats, loaded with articles of various kinds, producing necessaries for home consumption, and merchandizes for exportation, at an easy freight; which at present the settlers and proprietors on that river are deterred from improving, on account of their remote situation, and the length of a tedious land carriage; the expence of which amounts to at least thirty or forty per cent."[6]

Cook estimated that in one year the financial benefits of such a canal could equal the cost of construction.[7] This interest in a canal between the two rivers led Henry Mouzon, surveyor and mapmaker,[8] to produce a map of five routes for such a canal. Mouzon's map was drawn as early as 1775.[9] The five routes Mouzon identified were described in the *Army Report of 1881:* "First from Greenland Swamp through the north branch of Biggin Swamp to Biggin's Bridge, a distance of 23 miles; second, from Greenland Swamp, through the south branch of Biggin Swamp to Biggin's Bridge, a distance of 22 miles; and third, from Mattasee Lake to Wadboo Bridge, 16 miles; these three connected with the west branch of the Cooper River.

The fourth route was from Savannah Creek to Huger's Bridge, 14 miles and 40 chains [2,640 feet], and the fifth from Santee Creek to the same bridge on the east branch of the Cooper, a distance of 15 miles and 40 chains [2, 640 feet, for a total of 15.5 miles]"[10] (see fig. 2.1).

Following the end of the Revolution in 1783, interest in the canal increased. A meeting was held in Charleston on November 10, 1785: "to take into consideration the proposed plan of opening a communication by locks between Cooper and Santee rivers. The result of the meeting was to petition the Legislature at their next sitting, for a charter,—That the company should consist of a thousand shares at 100£ sterling each share, three guineas are to be paid down at subscribing, the remainder as shall be agreed upon when the charter is obtained."[11] There was considerable enthusiasm in Charleston for such an undertaking. The newspaper stated

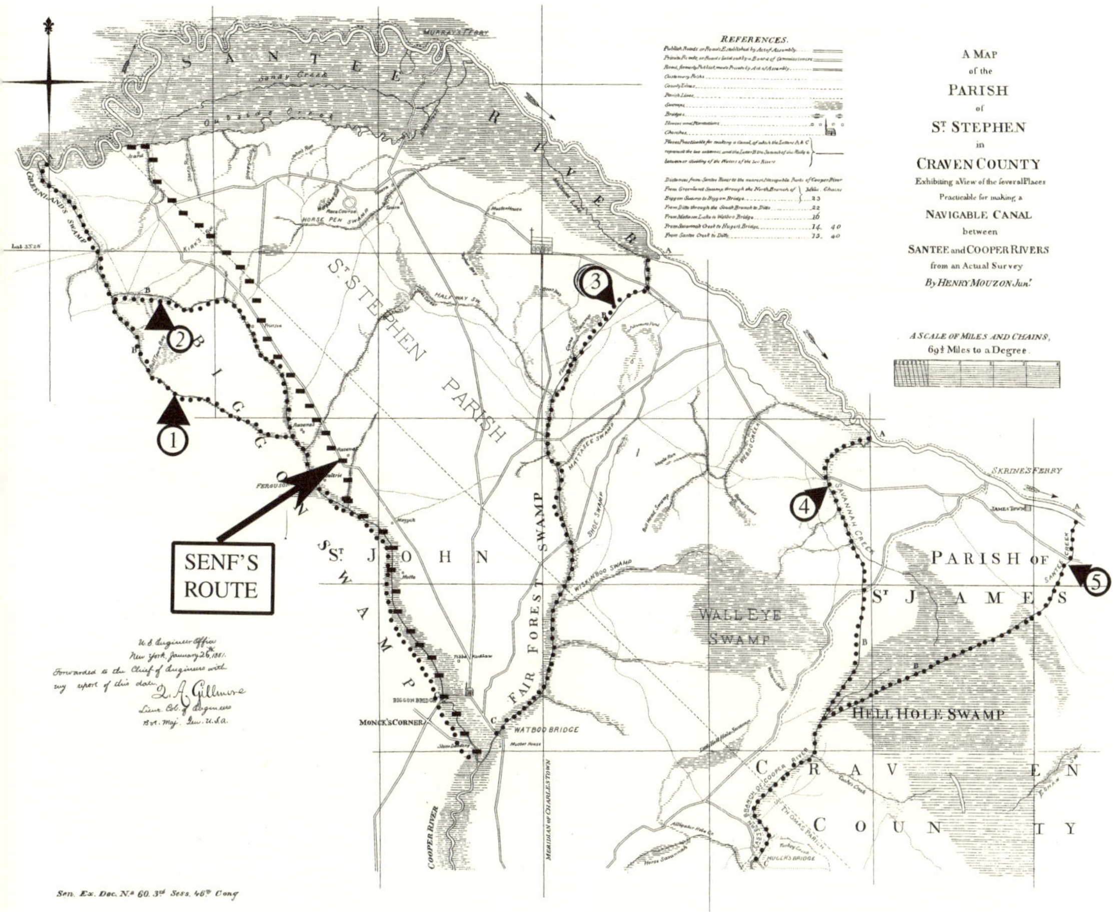

Fig. 2.1. John Christian Senf's route and Henry Mouzon's routes for the Santee Canal. Annotations by Sarah Prestemon using Mouzon's map from U.S. Senate, Ex. Doc. no. 60, *Letter from the Secretary of War* (1881), as a base

that "the above proposed plan is so highly approved of, that one gentleman has subscribed his name for three hundred shares."[12]

On February 7, 1786, a number of South Carolinians presented to John Lloyd, president of the South Carolina Senate, and to the Senate as a whole a petition calling on the legislature to charter a company to build the canal between the Cooper to Santee Rivers.[13] The petitioners pointed out "That a Considerable Number of Shares are already subscribed for, and the Whole Subscription will soon be filled, if the undertaking shall receive the Sanction of the Legislature, which is necessary to Carry it into execution."[14]

Six weeks later, on March 22, 1786, the South Carolina legislature passed "An Act to Establish a Company for the Inland Navigation from Santee to Cooper River."[15] The official name of the company was to be "The Company for the Inland Navigation from Santee to Cooper River," although it was usually referred to as just the Santee Canal Company. The company was given the right to levy tolls; the power to purchase needed land, not to exceed one thousand acres; the right to establish a ferry across the Santee at the north end of the canal; and other rights and privileges necessary to operate a canal.[16] The following day, March 23, 1786, the Santee Canal Company held their first meeting and elected Governor William Moultrie company president and John Rutledge vice president.[17]

As a surveyor, Mouzon could have laid out routes for the new canal, he probably did not have the expertise to design and construct the needed canal locks.[18] The new company needed a qualified civil engineer experienced in the design and construction of canals. At the time there were very few such engineers in the country. In fact few Americans had ever seen a canal lock.

In April 1786, two weeks after the passage of the Act of Incorporation, Governor Moultrie wrote to George Washington to see if they could get canal engineer James Brindley[19] to come to South Carolina and oversee the construction of the Santee Canal:

> Many of the Gentlemen who compose the Board of Directors [of the Company for the Inland Navigation from Santee to Cooper River] for carrying the plan into execution, having the honor of being personally acquainted with you were induced to recommend that, endeavour be made to obtain Mr. Brinley to superintend it, if in your Opinion, you judged him capable of the undertaking—it is said Mr. Brinley was constantly with his Brother [Brindley's uncle, the English canal engineer], while carrying on the Duke of Bridgewater's Works, & possessed great knowledge and abilities—To begin right, is all in all with us—the practicability of bringing it to perfection cannot admit of a doubt.
>
> However, the engagements that Mr. Brinley may now have of the same kind, may possibly prevent him entering a New One, yet if he could be spared only to inspect into the situation, & to give his opinion and directions how to proceed, it may at least prevent us from beginning wrong, & we may be going on untill he or some

other as equal can be procured: the Board of Directors will chearfully pay every expence that Mr. Brinley may be at, by coming here, exclusive of a compensation.[20]

This James Brindley was the nephew of the famous English canal engineer of the same name, who had built the first commercially successful canals in England, the Duke of Bridgewater's canals west of Manchester. The younger Brindley had immigrated to the United State in 1774 at the urging of John Ballendine[21] to work on the Potomac Canal, a pet project of George Washington's.[22]

From his experience in the Revolutionary War, Washington was familiar with Moultrie and the men who were members of the Santee Canal Company's board of directors[23] and would have been favorably disposed to their request (see fig. 2.2). On May 25, 1786, Washington responded to Governor Moultrie:

Mr. Brindley, nephew to the celebrated person of that name who conducted the work of the Duke of Bridgewater & planned many others in England, possesses, I presume, more *practical* knowledge of Cuts & Locks for the improvement of inland navigation, than any man among us, as he was an executive officer (he says) many years under his uncle in this particular business: but he is, I know, engaged with the Susquehanna company,[24] who are, I believe (for I saw Mr. Brindley about six weeks ago) in a critical part of their work. I have notwithstanding, written to a gentleman of my acquaintance who is not only a member of that Company, but one to whom

Fig. 2.2. Portraits of South Carolina Revolutionary War heroes, engraved by J. Rogers and published by Virtue and Yorston in the mid–nineteenth century. Generals Francis Marion, Thomas Pinckney, and Thomas Sumter all served on the first board of directors for the Santee Canal Company. General Andrew Pickens was active in establishing the Edisto Canal Company.

the business is chiefly confided, & near the spot, to know if Mr. Brindley's services can be dispensed with long enough to answer the purposes mentioned in your letter: his answer shall be forwarded as soon as it comes to my hands.[25]

Washington was able to arrange the loan of Brindley to the Santee Canal Company and on August 7, 1786, Governor Moultrie wrote to Brindley to arrange his trip to South Carolina:

By a letter from Mr. Hughes to General Washington, the Company for the Inland Navigation, are informed that the Susquehannah Company have given you permission to be absent for four months in the next winter, and that you have authorized Mr. Hughes to say that you will be in Charleston in December next. The Board of Directors have requested of me to urge your coming, and sooner if possible to South Carolina to take a view and give your opinion upon the probability of opening the communication between Santee and Cooper Rivers, they are perfectly satisfied with the knowledge and abilities you are possessed of in that branch of business and they are therefore unwilling to begin the great work, until you have surveyed and marked out the ground for the carrying of it into execution, and I assure you that every expense attending your coming here shall be paid you as also full compensation for your trouble, etc.[26]

We do not know exactly how long Brindley consulted on the Santee Canal in South Carolina, but we do know that he was in South Carolina by early 1787.[27] He left on March 17, 1787.[28]

Because Brindley was employed by the Susquehanna Canal Company, he could not accept a position from the Santee Canal Company. At this time John Christian Senf, a former Hessian soldier of Saxon birth, who was captured at the Battle of Saratoga during the Revolutionary War, was made chief engineer of the Santee Canal Company.[29] It is likely that Senf met Brindley and that the two consulted on the Santee Canal construction, particularly on its siting and the construction of its masonry locks.[30] It is also possible that Senf visited Brindley in Philadelphia in 1792, when Senf was in that city to sell shares in the Catawba and Wateree Navigation Company (also known as the Catawba Company).[31]

Senf's first task was to select a route for the Santee Canal. After reviewing the routes Henry Mouzon had surveyed earlier, Senf decided not to use any of Mouzon's five routes, a decision that later was considerably criticized. While he was in South Carolina, Brindley may have been involved in laying out the route that was used. In his August 7, 1786, letter to Brindley, Governor Moultrie had suggested that would be one of Brindley's tasks when he visited South Carolina. We do not know if Brindley did lay out a route for the canal, or, if he did, that it was the one Senf adopted. It is more likely that Senf was forced to choose the final canal route (see color plate 3)—from the northern terminus at White Oak Landing (also called

Porchers Embankment) on the Santee River to the southern terminus at Stoney Landing on Biggin Creek (west of Biggins Church)—because of local pressure from landowners desiring to have the canal pass through their lands. Benjamin Henry Latrobe,[32] probably the most capable architect-engineer in the United States at that time, wrote in 1811: "The Santee Canal indeed is the work of a German Colonel Senf, but he was forced to adopt a line he disapproved [of] by persons who had local interests. Local interests also ruined part of the Works of Weston[33] and both Senf and Weston bear unmerited blame."[34] Senf was later criticized by historian Frederick A. Porcher and others for selecting the canal route from White Oak Landing to Stoney Landing. Porcher suggested that, if a different route had been chosen, the canal would not have dried up during the droughts of 1816 and 1818–19. Other studies have indicated that any of the routes would have dried up during these drought years (see fig. 2.1).

Construction of the Santee Canal

Construction of the Santee Canal got off to a slow start. From the Act of Incorporation in 1786 until 1793, nothing was built. Because the records of the company have not survived, we have little direct knowledge of what was going on during these years. We know it was a period of discouragement for Senf. On January 3, 1790, he wrote to General Thomas Sumter: "I find my Zeal and few Abilities do not meet with that Incouragement I could, with honest pride, have expected—my too great confidence has exposed me to suffer Indignity—I thank God, that through the Interference of my friends I have been able to pary the Blow on my Reputation till this time."[35]

Construction of the canal finally began in 1793. There are not many surviving records detailing the construction of the Santee Canal. What we do know about its construction comes largely from Colonel Senf's final report, dated November 11, 1800,[36] dealing with the construction period of 1793 to 1800. To this report Senf attached letters discussing various aspects of the construction.[37] He also prepared a map, "General Plan of the Canal and its Environs between Santee and Cooper Rivers in the State of South Carolina."[38] This map was later copied, and the copy, but not the original, survives[39] (see fig. 2.3).

In addition to the few surviving company records, there are contemporaneous descriptions of the construction, including entries and sketches in the diary of Dr. Charles Drayton[40] (see figs. 2.4 and 2.5). There are also a few contemporary newspaper articles.[41] Later, more informative sources are the records of investigations undertaken by the South Carolina superintendent of public works in 1823[42] and by the Department of the Army in 1881.[43] In recent years there have been several archeological investigations on the remains of the Santee Canal.[44]

The most valuable recent report on the Santee Canal is the unpublished "History of the Santee Canal, 1785–1939," by Robert B. Bennett Jr. and Katherine H.

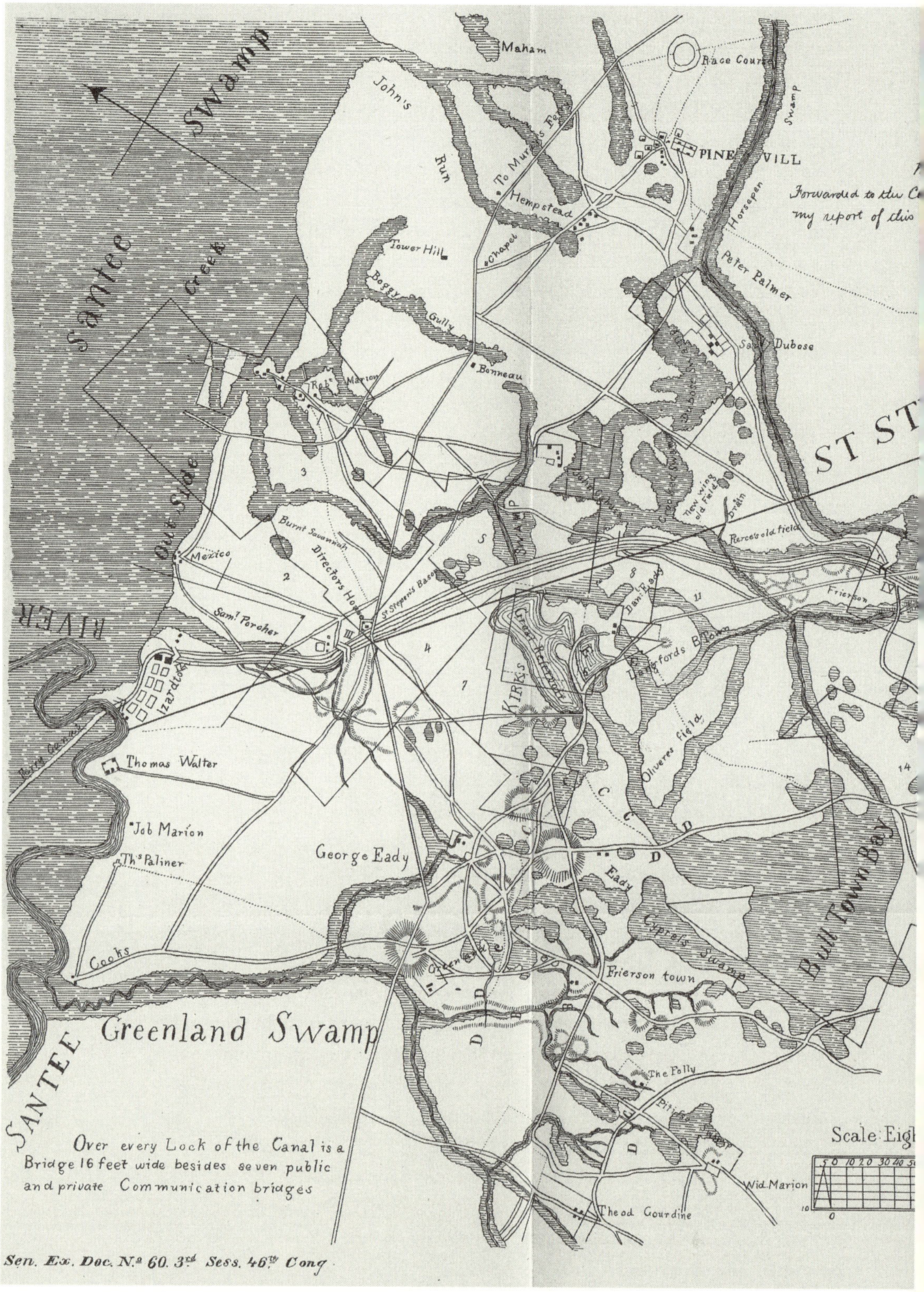

Fig. 2.3. *Above, facing, and overleaf:* John Christian Senf's map of the Santee Canal, as copied by J. P. Gaillard. From U.S. Senate, Ex. Doc. no. 60, *Letter from the Secretary of War* (1881)

GENERAL PLAN
of the
CANAL and its ENVIRONS between
SANTEE and COOPER RIVERS
in the
STATE of SOUTH CAROLINA
Commenced in the Year 1793 and finished in the Year 1800 by Christian Senf Colonel Engineer and Director in Chief of the Canal
U.S. Engineer Office.
New York, January 26th 1881.
Chief of Engineers with
o date
Q. A. Gillmore
Lieut. Col. of Engineers
Bvt. Maj. Gen. U.S.A.
PA
TEPHEN
Paul Ravenel
Rene Ravenel
Dan: Ravenel
Edw. Edwards
Peter Roy
Nath Marion
Chelsea
Northampton
St Johns Bason
Southampton
Rice Machines
Rice Field
Summertown
Saw Mill
Stephen Ravenel
Henry Ravenel
Geo: Porcher
Bl. White
Thos Porcher
Woodboo
ghty chains or One mile to One inch.
Miles

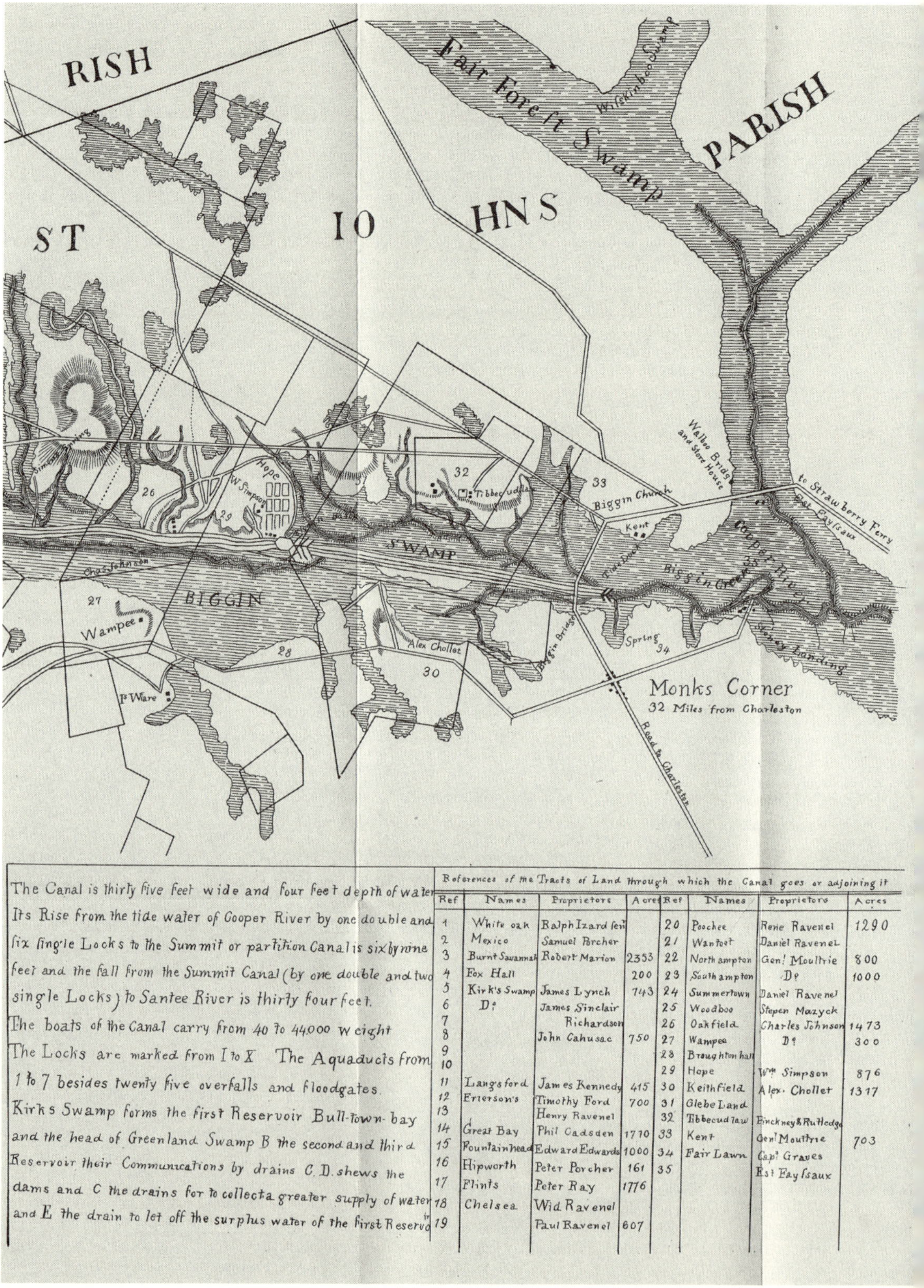

The Canal is thirty five feet wide and four feet depth of water
Its Rise from the tide water of Cooper River by one double and
six single Locks to the Summit or partition Canal is sixty nine
feet and the fall from the Summit Canal (by one double and two
single Locks) to Santee River is thirty four feet.
The boats of the Canal carry from 40 to 44000 weight
The Locks are marked from I to X The Aquaducts from
1 to 7 besides twenty five overfalls and floodgates.
Kirks Swamp forms the first Reservoir Bull-town-bay
and the head of Greenland Swamp B the second and third
Reservoir their Communications by drains C. D. shews the
dams and C the drains for to collect a greater supply of water
and E the drain to let off the surplus water of the first Reservoir

References of the Tracts of Land through which the Canal goes or adjoining it

Ref	Names	Proprietors	Acres	Ref	Names	Proprietors	Acres
1	White oak	Ralph Izard sen		20	Poochee	Rene Ravenel	1290
2	Mexico	Samuel Porcher		21	Wantoot	Daniel Ravenel	
3	Burnt Savannah	Robert Marion	2353	22	Northampton	Genl Moultrie	800
4	Fox Hall		200	23	Southampton	Do	1000
5	Kirk's Swamp	James Lynch	743	24	Summertown	Daniel Ravenel	
6	Do	James Sinclair		25	Woodboo	Stepen Mazyck	
7		Richardson		26	Oakfield	Charles Johnson	1473
8		John Cahusac	750	27	Wampee	Do	300
9				28	Braughton hall		
10				29	Hope	Wm Simpson	876
11	Langsford	James Kennedy	415	30	Keithfield	Alex. Chollet	1317
12	Frierson's	Timothy Ford	700	31	Glebe Land		
13		Henry Ravenel		32	Tibbecudlaw	Pinckney & Rutledge	
14	Great Bay	Phil Cadsden	1710	33	Kent	Genl Moultrie	703
15	Fountainhead	Edward Edwards	1000	34	Fair Lawn	Capt Graves	
16	Hipworth	Peter Porcher	161	35		Est Faysaux	
17	Flints	Peter Ray	1776				
18	Chelsea	Wid Ravenel					
19		Paul Ravenel	607				

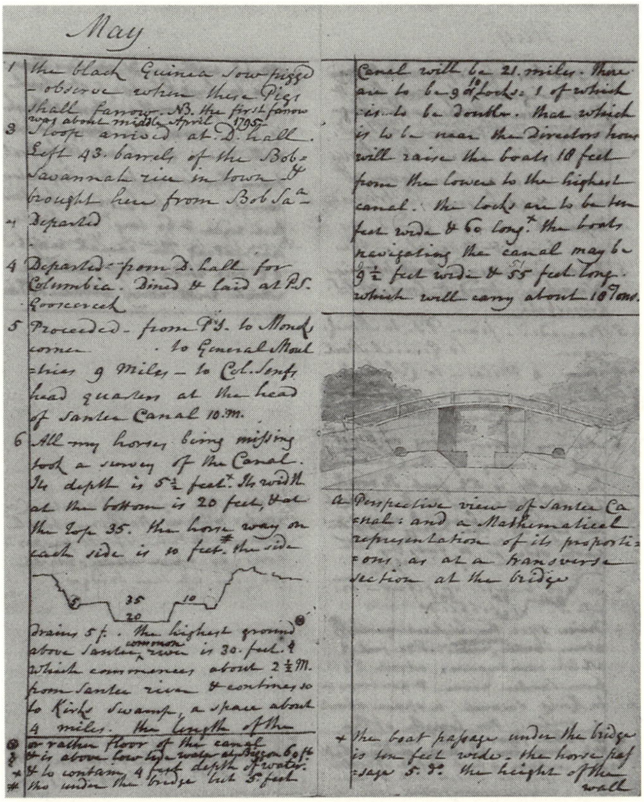

May

1 the black Guinea Sow pigged
- observe when these Pigs
shall farrow. NB. the first farrow
was about middle April 1795.
3 Sloop arrived at D. hall.
Left 43. barrels of the Bob
Savannah rice in town &
brought here from Bob Sa^a^ –
4 Departed

4 Departed from D. hall for
Columbia. Dined & laid at P.S.
Goosecreek
5 Proceeded – from P.S. to Moncks
corner . to General Moul
=tries 9 Miles – to Col. Senfs
head quarters at the head
of Santee Canal 10.M.
6 All my horses being missing
took a survey of the Canal.
Its depth is 5½ feet.* Its width
at the bottom is 20 feet, & at
the top 35. the horse way on
each side is 10 feet.# the side

35 10
20

drains 5 f. the highest ground
above Santee common river is 30. feet.
which commences about 2½ M.
from Santee river & continues to
to Kirks Swamp, a space about
4 miles. the length of the

or rather floor of the canal
is above low tide water at the Biggon 6 ft.
* to contain 4 feet depth of water.
# tho under the bridge but 5. feet

Canal will be 21. miles. There
are to be 9 or 10 locks. 1 of which
is to be double. that which
is to be near the Directors house
will raise the boats 10 feet
from the lowest to the highest
canal. the locks are to be ten
feet wide & 60 long.* the boats
navigating the canal may be
9½ feet wide & 55 feet long.
which will carry about 10 Tons.

A Perspective view of Santee Ca
=nal: and a Mathematical
representation of its proporti
=ons as at a transverse
section at the bridge

* the boat passage under the bridge
is ten feet wide. the horse pas
=sage 5. do. the height of the
wall

Fig. 2.4. Excerpt from the diary of Dr. Charles Drayton, including a May 5, 1795, description of the Santee Canal and its construction. From the collections of Drayton Hall, National Trust for Historic Preservation, Charleston, S.C. Courtesy of Drayton Hall

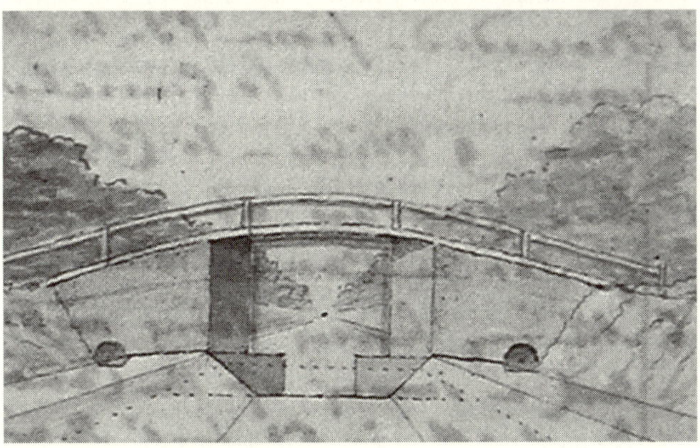

Fig. 2.5. Detail from Drayton's May 5, 1795, diary entry, a perspective drawing of the Santee Canal showing "its proportions as at a transverse section at the bridge." From the collections of Drayton Hall, National Trust for Historic Preservation, Charleston, S.C. Courtesy of Drayton Hall

Richardson, written as background for the development of Old Santee Canal State Park at Moncks Corner, South Carolina.[45] Also valuable is Porcher's *The History of the Santee Canal,* written in 1875, published in 1903, and reprinted several times since.[46] There are a few other sources about the Santee Canal, but they are not extensive. The primary source for the history of its design and construction is Senf's own account.

Senf's first problem was how to build a summit-level canal that could surmount the ridge dividing the Cooper and Santee watersheds. This ridge was approximately seventy feet high above tidewater of the Cooper River and about thirty-five feet higher than the Santee River. The key to the success of this canal—as with all summit-level canals—would have been to site the canal so that there was adequate water at the summit level during all seasons of the year.

Senf began his reconnaissance of the area in 1786 and provided his design recommendations to the president and board of directors by January 16, 1787, and again on April 2, 1787.[47] Senf had to develop these recommendations without extensive meteorological and hydrological data. A modern civil engineer designing a canal like the Santee Canal would use water-flow measurements taken at various times of the principal water courses so as to determine if there would be adequate water flow for the proposed canal. Because river flow can fluctuate greatly from season to season and from year to year, this data is now typically gathered over long periods. Senf did not have the benefit of this sort of information and could not observe water flow over time because he needed to finish his reconnaissance and develop his recommendations before the legislature adjourned. He had to rely on his own observations and on nearby residents' anecdotal recollections about past droughts and floods.[48]

Senf's recommended starting point, White Oak Landing on the Santee River, is immediately east of Greenland Swamp, which was the recommended starting point for two of the routes identified by Henry Mouzon. At this point the Santee River was normally 270 feet wide and 18–20 feet deep.[49] Senf's route began at the lands of Ralph Izard, and that route was 5 feet higher than Mouzon's proposed route through Greenland Swamp (see fig. 2.6, upper center). Despite the higher elevation of the canal at its entry to the Santee River, Senf was convinced that this route would have adequate water at the summit level. He also felt that this location would protect the canal from freshets from the Santee River.

In order to save water needed at the summit level, Senf proposed a canal with diminutive dimensions. The canal was to be thirty-two feet wide at the surface and four feet deep with three feet of water.[50] This was later increased to thirty-five feet wide at the surface, thirty-two feet width at the bottom and five and a half feet deep with four feet of water.[51] The boats to travel on this canal were to be fifty feet long and nine feet wide. The locks were nine and a half to ten feet wide with sixty feet between the lock gates.[52] Some locks were later extended, generally to sixty-six feet, to allow longer boats pass through.[53] The locks were constructed of brick made from clay in the area that was burned on location.

In April 1787 Senf submitted his estimate for this work: £55,620, or about $233,000—an immense sum for South Carolina in the economically depressed years immediately following the Revolution:

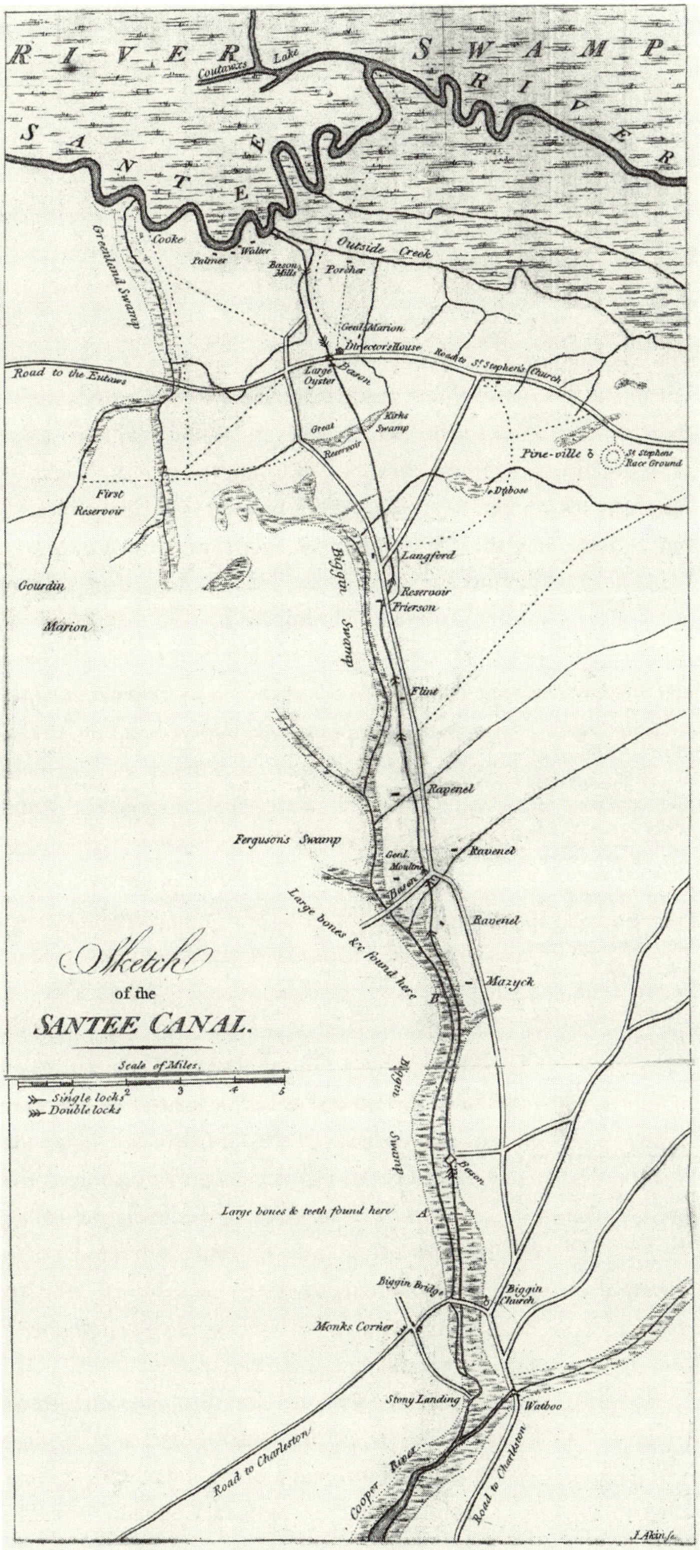

Fig. 2.6. Map of the Santee Canal, from John Drayton, *A View of South-Carolina* (1802)

1st.	For cutting the canal 21 miles long and 32 feet wide, to be, on an average, 6 feet deep, allowing 12 pence a cubic yard, including tools	£34,482
2d.	For building substantial locks, the dimensions and falls as above, allowing, as usual £100 to a foot fall	£11,000
3d.	For the drain from the reserve B, of 2 miles length, 10 feet deep, on an average, and 24 feet at the top	£2,738
4th.	For 8 or 9 aqueducts (i.e. culverts) and draw-bridges	£4,400
5th	For other small drains, dams, fencing, etc.	£3,000
		£55,620[54]

This amount was not totally unexpected because the capitalization envisioned by the company's stock solicitation in 1786 was for one thousand shares at £100 each.

Work did not begin on the canal until six years later, in May 1793. Probably this delay was owing to a lack of money. It is likely that the company had to turn to northern and overseas investors to sell its shares. Senf traveled to Philadelphia in 1792 to solicit investment in the Santee Canal as well as in the Catawba Company, another early South Carolina canal company.[55]

Financing began to come together for the Santee Canal Company in 1791. A meeting concerning "Matters of Importance" was held on April 12, 1791.[56] A second meeting was called for October 17 of the same year to discuss, "a most flattering prospect . . . for the beginning and carrying to full effect, the intended canal between the Santee and Cooper rivers."[57]

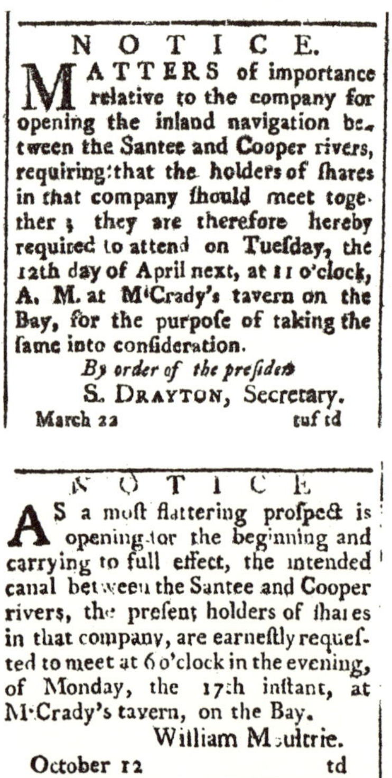

NOTICE.

MATTERS of importance relative to the company for opening the inland navigation between the Santee and Cooper rivers, requiring that the holders of ſhares in that company ſhould meet together; they are therefore hereby required to attend on Tueſday, the 12th day of April next, at 11 o'clock, A. M. at M'Crady's tavern on the Bay, for the purpoſe of taking the ſame into conſideration.

By order of the preſident
S. Drayton, Secretary.
March 22 tuf td

NOTICE.

AS a moſt flattering proſpect is opening for the beginning and carrying to full effect, the intended canal between the Santee and Cooper rivers, the preſent holders of ſhares in that company, are earneſtly requeſted to meet at 6 o'clock in the evening, of Monday, the 17th inſtant, at M'Crady's tavern, on the Bay.

William Moultrie.
October 12 td

City Gazette, April 1, 1791, and October 13, 1791

What these "Matters of Importance" were or what the "flattering prospect" was not recorded, but by 1792 the company had begun to get organized. At their January 10, 1792, meeting, the stockholders adopted rules for the company (see appendix). The books were opened for share subscriptions on May 1, 1792, by David Ramsay, Thomas Jones, and Nathaniel Russell. The cost to subscribe to one share was £5 ($21.45). Under the rules of the company, the subscriber agreed to pay an additional £5 each January and July thereafter until the canal was completed. In addition the company president and directors had the authority to assess an additional amount up to £5 twice a year. Failure to meet any of these payments resulted in the loss of the share, funds already paid into the company, and any interest owed.[58] The yearly cost of each share was therefore £10 ($42.90) per year and sometimes more. At that time laborers were being paid fifty cents a day and artisans a dollar a day, so it is clear that the subscribers of the Santee Canal Company were from the wealthier classes—the merchants and the planters. Theoretically one thousand shares at £10 per year over eight years (1793–1800) would have generated £80,000—plus £5,000 from the initial payments required from every new subscriber and another £5,000 from the extra payment ordered in 1799, to be paid on the third Monday of January, 1800.[59] This would have yielded a total possible capitalization of £90,000—well in excess of the £55,620 (approximately $233,000) cost estimated by Senf. But the company had continuing difficulty meeting its expenses. Costs were much higher than Senf predicted. In addition the company was unable to sell all one thousand shares. To fill the gap, the board of directors authorized a number of lotteries, and they also borrowed money to complete the canal.[60]

By order of the Board of Directors, the following Rules of the Santee Canal Company, are publiſhed, to wit:

RULE 10th.

THAT every proprietor ſhall pay five pounds ſterling on each ſhare on the third Monday in January 1793, and the like ſum on the third Monday in July 1793, and on the third Mondays in January and July, in every ſucceeding year, until the Canal ſhall be completed: but if the Preſident and a majority of the directors ſhall judge it neceſſary to raiſe a larger ſum, they ſhall be at liberty to call for the ſame, to be paid on notice of ſix months; provided, that the ſum ſo to be called for on each ſhare, ſhall not exceed five pounds; and provided alſo, that the requiſitions ſhall be made on the third Mondays in January or July.

Rule 11th. That if any proprietor ſhall fail to make any payments, to which by the rules of the company he or ſhe may be liable when it ſhall become due, or within three days thereafter, every ſuch proprietor ſhall forfeit to the company all the preceding payments, and alſo his or her intereſt in the ſaid company.

Payments will be received by Nathaniel Ruſſell, No. 16, Bay, or Thomas Jones at the Bank of South Carolina; and it will be requiſite for thoſe concerned, to bring their ſcripts, that receipts may be endorſed thereon.

Charleſton, 10th December, 1792.

S. DRAYTON, *Secretary.*

N. B. The ſcripts in the names of the following perſons, remain as yet in the ſecretary's hands, to be delivered when applied for, to wit:

Francis Marion,
James Graham,
Ephraim Mitchell,
Blakelea White,

December 11. mth td

City Gazette, December 9, 1792

Description of the Santee Canal

Senf's canal had two separate and distinct sections. The northern section ran from the inlet at the Santee River to the summit level, and the southern section took the canal down from the summit level to the outlet at the Cooper River.

Northern Section (Santee River to Lock 6)

The northern section began at the Santee River. In addition to building a canal to the Cooper River, the company had also been authorized to initiate a ferry service across the Santee River. At the canal's entrance Senf constructed a channel one hundred feet wide to allow ferry boats to lie in the canal and still permit canal traffic to pass.

On Tueſday, 13th inſtant,

Before my ſtore near the Exchange, will be ſold,

Fifty Canal Scripts,

In lots from one to ten. Any perſon willing to purchaſe in the mean time may apply to

Job Colcock.

November 9. td

City Gazette, November 9, 1792

Across from the entrance to the canal, on the north side of the Santee River, was a swamp five and a half miles wide (see fig. 2.6, upper left). To initiate the ferry service across the river, Senf had to construct a two-and-a-half-mile channel through this swamp to Cotteaux Lake. Dug during the low-water summer months, this channel was forty feet wide and two to three feet deep. The lake was used as a slack-water passage for three miles. The northern landing for the ferry was established between Nelson's Blue House and Norvell's, three miles above the Santee River. This ferry was intended to draw commerce from the Black River and Lynches River (then referred to as Lynche's Creek) to the Santee Canal. Senf suggested that the company build warehouses at the ferry-boat landing at the entrance to the Santee Canal, on the south side of the river[61] (see fig. 2.7, far left).

At the Santee River entrance to the canal, Senf had built what he called "epis," probably rock cribs intended to divert the river away from the canal entrance at

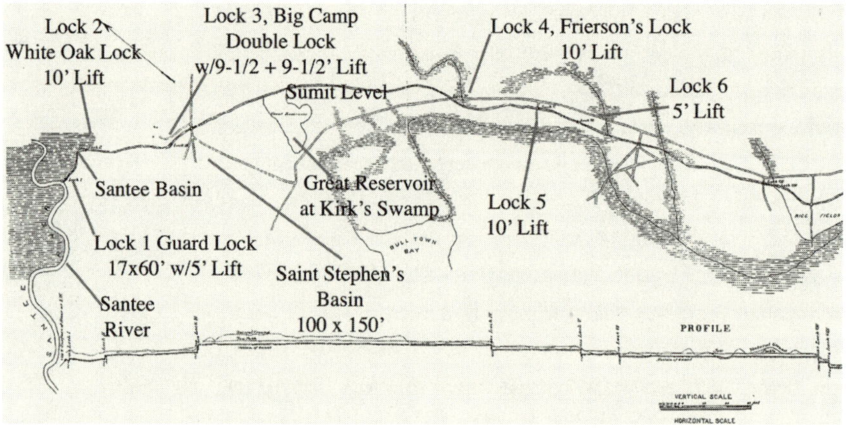

Fig. 2.7. Plan and profile of the northern section of the Santee Canal, Santee River to lock 6, from "Chart of Survey of the Santee Canal, S.C." From U.S. Senate, Ex. Doc. no. 60, *Letter from the Secretary of War* (1881), with annotations by Sarah Prestemon

White Oak Landing so as to halt bank erosion.[62] Later, during the drought of 1818–19, a wooden wing dam was added to swell the water level at the canal entrance to allow the boats to cross the sill of the guard lock.[63]

To ascend from the Santee River to the five-mile-long summit level (level 3–4), Senf built three locks. The first, located 150 feet from the Santee River, was the guard lock with a lift of 5 feet and with an extra wide chamber (17 by 60 feet),[64] which enabled boats and rafts to ascend the Santee Canal to the Santee basin, where cargo could be loaded onto the narrow canal boats. This guard lock (lock 1) protected the canal from floods on the Santee and, when the river was low, protected the canal from outflow to the lower river level. Like most of the eleven locks Senf built on the Santee Canal, lock 1 was constructed of brick with wooden gates. Senf did not describe the foundations he built for this and the other locks, but they were probably either wood pilings or rafts of timbers. On top of this foundation, Senf would have constructed a wooden structure, or grillage, and on top of the grillage a brick floor.

Bricks used to build the lock walls were most likely burnt locally.[65] Recent archeological evidence indicates that at least one of the Santee Canal's eleven lock floors was paved with bricks.[66] Probably they all were. The brick would have been burned on the site from clay dug nearby. Senf described the excellent clay that was available locally. The hollow quoins, in which the turning posts of the lock gates were to be mounted, were made of stone.[67] It is said that the wooden gates, flooring, and other wooden members were made of cypress from the Santee Swamp.[68]

The brick lock walls Senf built acted as retaining walls and needed to be strong enough to hold back the earth behind them. Recent photographs of the remnants of these locks, taken during drought conditions on Lake Moultrie (see color plate 4) indicate that Senf built counterforts on the inside or land side of the lock walls to strengthen them. Counterforts look like buttresses, but they function in tension to hold back the walls, rather than in compression as buttresses normally do.[69]

The mortar for the brick locks could have been produced from burned limestone or oyster shells. Although Senf did not use hydraulic cement in these locks to resist erosion of the mortar (as was done in later canals), the locks held up well. Seventy-five years after they were built, when they were examined by U.S. Army engineers, the locks were still in fair condition.[70] Across the guard lock and each of the other ten locks, Senf constructed a sixteen-foot-wide wooden bridge with adequate clearance for the canal boats below.[71]

Level 1–2. The level of the water between locks was named for the locks at either end. So the first level of water was called level 1–2. Above lock 1, the guard lock, the canal prism (or canal section) narrowed from 100 feet to 40 feet wide at the surface with water 6 feet deep.[72] At 1,000 feet beyond lock 1 on level 1–2, Senf built a small turning basin of 85 by 350 feet, a wide area in the canal permitting boats to turn around.[73] Beyond the turning basin, the canal was shaped like a half moon, 1,000 feet in length and 245 feet wide in the middle. At the end of this arc the canal reached the Santee Basin and lock 2, known as White Oak Lock. Like lock 1, lock 2 was of brick and stonemasonry construction, 10 feet wide and 60 feet between the gates but with a 10-foot lift.[74] At the bridge over lock 2, the town of Izardton—named for Ralph Izard, one of the landowners and a director of the Company for the Inland Navigation from Santee to Cooper Rivers—was developed.[75] Senf also developed a mill site there with a 4-foot wide sluice, a 6-foot fall, and a brick-paved waste way to eliminate scour, the erosion of the soil by water action. Immediately beyond lock 2 Senf built a small, 30- by 60-foot basin as a dry dock for canal boats. This dry dock could service two boats at one time.[76]

Level 2–3. Immediately above lock 2, on level 2–3, the canal prism narrowed again, to 35 feet wide at the top and 20 feet wide at the bottom with a water depth of 4 feet.[77] At this point the canal was carried on an embankment approximately 3 feet above the natural ground. Level 2–3 was 1 mile and 2,957 feet (1.56 miles) long. This level was supplied with water from a reservoir in Greenland Swamp immediately west of the canal. This reservoir failed during the drought years of 1816 and 1818–19.[78] Lock 3 marked the entrance to the summit level of the canal—its highest elevation.

Level 3–4. The summit level, level 3–4, began at lock 3, known as Big Camp. Lock 3 was a double lock, with each having a 9½-foot lift (for a total lift of 19 feet). Lock 3 was also built of brick and stonemasonry and like lock 2 was 10 feet wide with 60 feet between the gates. Senf equipped lock 3 with a side chamber designed to save half the water that would otherwise have been expended down the canal during the transit of a canal boat. When a canal boat went downstream through lock 3, this innovation made half the lock's water flow into the side chamber and held it there. When the side chamber was full, the remaining water in the lock was discharged down the canal. The downstream gates were then opened to permit the canal boat to progress down the canal. When a boat was lifted by the lock, that boat would proceed through the downstream gates, and the gates closed behind it. The water in the side chamber was then be released into the lock basin, and a deficiency of water was met by opening the valves in the upstream lock gates.

Adjacent and just above lock 3 was the bridge carrying the road to Eutaws (see fig. 2.6).[79] Above this bridge Senf excavated St. Stephen's Basin (100 by 150 feet), used for loading and unloading cargoes and permitting canal boats to turn around. This was the headquarters area of the canal, where Senf constructed a two-story company warehouse, the company's work yard, the canal company director's house, and a lock house for the keeper of lock 3.[80]

Water for the summit level and adjacent levels came from the Great Reservoir at Kirk's Swamp west of the canal, designated as the first reservoir, and from three other reservoirs. The Great Reservoir was 1,200 feet wide.[81] Other reservoirs for this level were from the head of Greenland Swamp and by the swamp called Bull Town Bay.[82] Senf built a fourth reservoir at the south end of the summit level, at the head of Biggin Swamp near lock 4.

The summit level was the location of the deepest digging on the Santee Canal. Excavation extended 14 to 16 feet below grade. The width of this cut at the surface was 110 feet,[83] which was required to accommodate the 35-five foot width of the canal at its surface, the 10-foot width of the towpaths on each side of the canal, and the necessary berms and ditches.

The summit level was almost 5 miles long.[84] At the southern end of the summit level was lock 4, known as Frierson's Lock. This lock was 10 feet wide and 60 feet long with a 10-foot lift.[85] Like lock 3, Frierson's Lock was equipped with a chamber that would save half of the water in the lock when a boat was passed through.[86] South of this lock the Santee Canal began to parallel Biggin Swamp (see figs. 2.8 and 2.9).

Level 4–5. Immediately south of Frierson's Lock was the level between it and lock 5. Level 4–5 was 122 chains or 1 mile and 2,772 feet (1.525 miles) long.[87] This level had no reservoir or spring to supply its water. Its only source was water discharged through the summit level lock 4 when it passed a boat. Consequently this level was first on the Santee Canal to go dry during the droughts of 1816 and 1818–19.

Lock 5 was also 10 feet wide with 60 feet between gates and a 10-foot lift.[88] Like the other locks, it was constructed of brick with stone hollow quoins for the gate hinges.

Level 5–6. The level between lock 5 and lock 6 was 52 chains 80 links (3,484.8 feet or 0.66 miles) long and was fed with water from Frierson's Swamp.[89] This level crossed a ravine 400 feet in width. Under the canal prism was the second of seven "aqueducts" (that is, culverts), semicircular masonry structures, each with a base 2 feet wide and a rise of 4 feet in the top of the arch.

Fig. 2.8. Remains of Frierson's Lock (lock 4), Santee Canal, 1930s. Photographer unidentified. Historic American Buildings Survey, Prints and Photographs Division, courtesy of the Library of Congress. The sign on the tree reads, "Frierson's Lock . . . Built 1800."

Fig. 2.9. Another view of the remains of Frierson's Lock, 1930s. Photographer unidentified. Historic American Buildings Survey, Prints and Photographs Division, courtesy of the Library of Congress

At the end of level 5–6, lock 6 was also 10 feet wide with 60 feet between gates, and it had a lift of 5 feet.[90] It was also constructed of brick with stone hollow quoins.

Level 6–7. The level between lock 6 and lock 7 was about three miles long and included Red Oak Swamp.[91] This level was not fed by springs or reservoirs and consequently went dry during times of drought, such as in 1816 and 1818–19.[92]

Southern Section—Lock 7 to Cooper River

The southern section of the Santee Canal began at lock 7, known as Black Oak Lock, which was also 10 feet wide with 60 feet between gates and had a 10-foot lift.[93] At

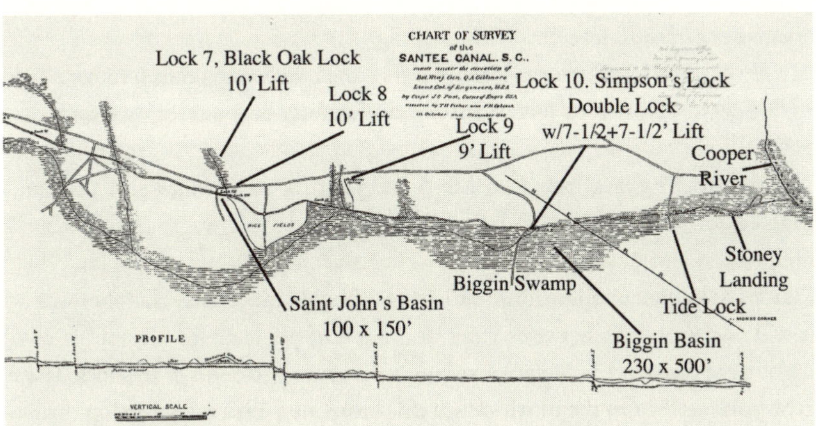

Fig. 2.10. Plan and profile of the southern section of Santee Canal, from "Chart of Survey of the Santee Canal, S.C." From U.S. Senate, Ex. Doc. no. 60, *Letter from the Secretary of War* (1881), with annotations by Sarah Prestemon

Fig. 2.11. Remains of Black Oak Lock (lock 7), Santee Canal, 1930s. Photographer unidentified. Historic American Buildings Survey, Prints and Photographs Division, courtesy of the Library of Congress

Fig. 2.12. Remains of a lockhouse at Black Oak Lock (lock 7), Santee Canal, 1930s. Photographer unidentified. Historic American Buildings Survey, Prints and Photographs Division, courtesy of the Library of Congress

the lower end of this lock, Senf constructed St. John's Basin, which was 100 by 150 feet, the same size as the St. Stephen's Basin at lock 3. Senf recommended that the company construct warehouses here. St. John's Road, a north-south road paralleling the Santee Canal, was 450 feet east of St. John's Basin. Senf had a bridge constructed over lock 7 so that St. John's Road could be connected with Moultrie's causeway over Biggin Swamp and from there to the main road located 3 miles west of the canal[94] (see figs. 2.10, 2.11, and 2.12).

Level 7–8. The level between lock 7 and lock 8, was 16 chains 10 links (1062.6 feet or 0.2 miles) long.[95] Lock 8 was another brick-and-stone structure 10 feet wide with 60 feet between gates and a 10-foot lift. Under this lock Senf had constructed two culverts, called aqueducts 4 and 5, each 4 feet wide with a rise of the semicircular arch of 4 feet.[96]

Level 8–9. The level between locks 8 and 9, was 1 mile and 2,640 feet (1.5 miles) long.[97] Lock 9 was also 10 feet wide with 60 feet between gates; it had a lift of 9 feet.[98]

Level 9–10. The level between lock 9 and lock 10 was 3 miles and 32 chains (2,112 feet or 3.4 miles) long. It was well supplied with water from Mazycks Spring.[99]

Lock 10, known as Simpson's Lock, was a double lock. Each chamber was 10 feet wide with 60 feet between gates and a lift of 7½ feet (for a total lift of 15 feet).[100] Senf constructed Biggin Basin immediately to the north of this lock. It was 230 by 500 feet.[101] On the north side of this basin, he built a two-story, 25- by 60-foot brick warehouse (see color plate 5).[102] On the west side he constructed a dry dock similar to the dry dock near Santee Basin.[103]

Level 10–Tide Lock. The distance from lock 10 to Tide Lock was 2 miles 45 chains (2,970 feet or 2.56 miles) long.[104] Senf specified that Tide Lock was to be the same size as the others with 60 feet between the gates.[105] Although all the other locks were built of brick and stone, recent archeological investigations have revealed that Tide Lock was constructed of wood.[106]

Tide Lock to Stoney Landing. From Tide Lock to Stoney Landing was one mile via Biggin Creek. Senf cut off a turn in the creek so as not to impede flood waters and thereby endanger the canal. From Stoney Landing to where Biggin Creek joins Fair Forest Creek—the beginning of the Cooper River—was an additional three-quarters of a mile of in-stream navigation.

In all Senf designed and oversaw construction of eleven locks, ten of brick and stone and one of wood; twenty-six bridges over the canal; seven brick culverts; twenty-one and a half miles of canal, including towpaths, puddling, basins, lock-keepers' houses, warehouses, and other structures; twenty-five flood gates and overfalls (sluices); and four reservoirs for supplying water to the canal.

Workers, Slaves, and the Building of Santee Canal

To build the Santee Canal, Senf needed labor and a lot of it. Because this construction took place in South Carolina during the 1790s, much of this labor was supplied by black slaves. South Carolina agriculture was in a depression during the 1790s because of the loss of the English bounty, or bonus, for crops such as indigo, so planters had many underutilized African American slaves who could be rented to Senf and the canal company. Advertisements for black slaves and other workers began to appear in the Charleston newspapers in August 1792:

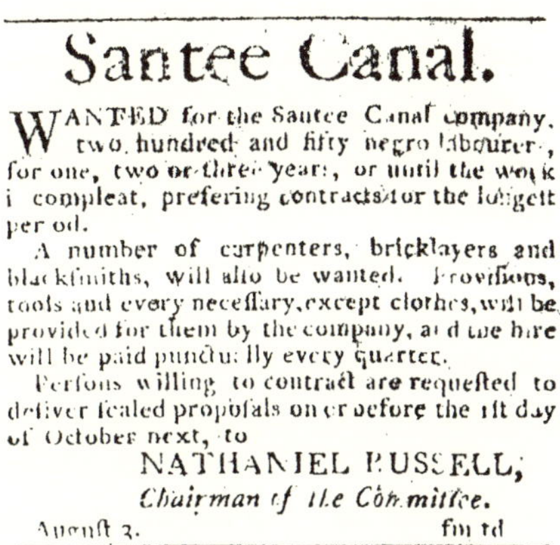

Santee Canal.

WANTED for the Santee Canal company, two hundred and fifty negro labourers, for one, two or three years, or until the work is compleat, prefering contracts for the longest period.

A number of carpenters, bricklayers and blackſmiths, will alſo be wanted. Proviſions, tools and every neceſſary, except clothes, will be provided for them by the company, and the hire will be paid punctually every quarter.

Perſons willing to contract are requeſted to deliver ſealed propoſals on or before the 1ſt day of October next, to

NATHANIEL RUSSELL,
Chairman of the Committee.

August 3. fm td

City Gazette, August 10, 1792

The planters wanted to rent both female and male slaves to the canal builders. Although women were not as strong, they were less expensive, and to have a sufficient workforce, Senf was forced to hire females as one third of his workers. The following advertisement first appeared in the Charleston newspapers in August 1792 and ran for several months thereafter:

Santee Canal.

WANTED for the Santee Canal company, two hundred and fifty negro labourers, for one, two or three years, or until the work is compleat, prefering contracts for the longest period.

A number of carpenters, bricklayers and blackfmiths, will alfo be wanted. Provifions, tools and every neceffary, except clothes, will be provided for them by the company, and the hire will be paid punctually every quarter.

Perfons willing to contract are requefted to deliver fealed propofals on or before the 1ft day of October next, to

NATHANIEL RUSSELL,
Chairman of the Committee.

Auguft 3. tin td

SANTEE CANAL.

WANTED for the Santee Canal Company, as foon as convenient after the 30th November next, two hundred and fifty negro labourers in a proportion not exceeding one third women to two thirds men, for one, two and three years, preferring contracts for the longeft periods; the hire at the rate of fifteen pounds per annum, will be paid quarterly. Tools, provifions, and every neceffary including medical affiftance (clothes excepted) will be found by the company. Propofals will alfo be received for a medical gentleman, to attend the aforefaid negroes by the year. Apply to

NATHANIEL RUSSELL,
Chairman of the Committee.

October 8. nuh td

City Gazette, August 3 and December 6, 1792

Although Senf initially had to take one third of the total workforce as women, he was later required to use women as half the workers.[107] The workforces on other canals, such as the Potomac Canal, had included women in jobs such as cooks, nurses, and even teamsters.[108] The Santee Canal was the only American canal on which women were used in large numbers as laborers.

At the beginning of construction the Santee Canal Company was paying slave owners £15 to £16 (about $64 to $69) per year for each slave. Toward the end of the construction, the Company was paying £24 (about $103) for men and £20 (about $86) for women per year.[109] There were several reasons for this increase. The global conflict between England and France raised demand for all sorts of commodities and increased prices. The spread and development of cotton plantations had increased the demand for slaves, which drove up the price for them. In addition to the amount paid to the slave owners, the company paid for the food, tools, and other items the slaves needed. This required an additional £15 to £18 pounds (about $64 to $77) per slave per year.

Senf began work on the canal in May 1793 with 10 laborers; by July 1793 that number had increased to 150; by the end of 1793, the total number had swollen to 1,000. For 1794 Senf employed 800 workers, and for 1795 he employed 700. This may sound like a lot of workers, but Senf pointed out that sick workers, runaways, and laborers who performed specialized tasks, substantially lowered the number of laborers available for excavation:

Suppose now seven hundred laborers had been employed per year from the beginning of 1793, the following deductions should of course be made from that number, viz.

4 Sick per Hundred 28
4 Runaways [per hundred] 28
Drivers 36

Cooks for Laborers and white Persons	34
Colliers, cart and waggon Boys, Butchers, etc.	12
Sawyers and squarers	16
Labourers to assist Tradesmen	30
[Laborers] for Brick yards	26
	200[110]

Therefore out of 700 workers, only 500 did excavation and embankment work.

In the early years of construction, Senf had an assistant engineer to help him, a Mr. Depesseville. Not much is known of Depesseville, but it is possible that he was a French émigré fleeing the French Revolution and subsequent wars. He remained on the Santee Canal until 1796 when he had to leave for health reasons.[111]

By July 1793 Senf and his workers had excavated twelve hundred feet of the twenty-two-mile-long canal. Perhaps not much of an accomplishment but the company seized on the results to issue a highly optimistic report on the canal that appeared in the Charleston newspapers:

CHARLESTON,

TUESDAY, *July* 2, 1793.

Yeſterday the ſtockholders in the Santee Canal Company had a meeting On enquiry the affairs of the company were found to be in a moſt proſperous ſituation. 195 laborers, beſides tradeſmen, were conſtantly employed, and proceeding with ſpirit. Upwards of 400 yards of the canal were completed. It farther appeared, that there were upwards of 21,000 dollars lodged in the banks belonging to the company and ſubject to its drafts; and that the tools, and upwards of 7000 buſhel of corn, were purchaſed and paid for. The hire of the laborers, and all other incidental expences, having been diſcharged, and there being no preſent occaſion for money, it was reſolved, that the ſtockholders ſhould only be called on for twenty ſhillings a ſhare, at the next day of payment.

Such is the flouriſhing ſtate of the company, and ſo flattering are its proſpects, that, it is to be hoped, the ſpeedy completion of the canal will enſure immenſe advantages both to the country at large, and the ſtockholders in particular.

City Gazette, July 2, 1793

Senf chose to supervise the canal building directly rather than hiring contractors and paying them an agreed upon unit sum of money per cubic yard excavated (the system later used in constructing the Erie Canal). This meant the company needed to hire overseers to supervise the work of the slaves. For every twenty slaves working on the canal, the company hired one overseer, who reported to and was paid by the company. In 1793 overseers were initially paid £30 per year (about

$129). By the end of canal construction in 1800 their pay had doubled to £60 (about $257) per year.

The slaves excavated the canal prism by shovel and transported the earth by wheelbarrow, usually on wooden boards laid for that purpose.[112] It was estimated that a slave could dig and transport two cubic yards of earth a day.[113] A day's labor was from sunrise to sunset with two hours off for a midday meal. A work week was six days with Sundays off.

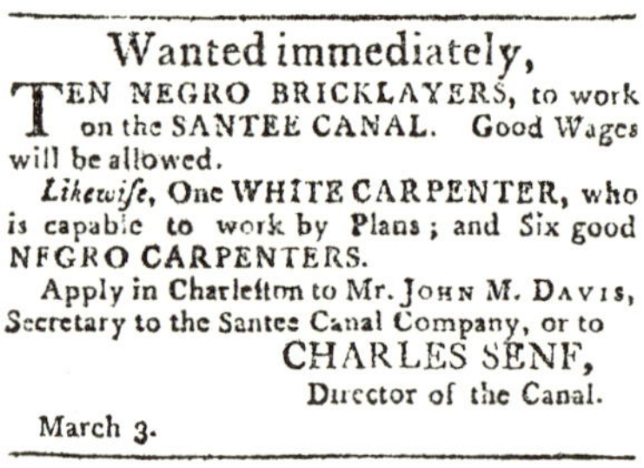

Wanted immediately,

TEN NEGRO BRICKLAYERS, to work on the SANTEE CANAL. Good Wages will be allowed.

Likewise, One WHITE CARPENTER, who is capable to work by Plans; and Six good NEGRO CARPENTERS.

Apply in Charleston to Mr. JOHN M. DAVIS, Secretary to the Santee Canal Company, or to CHARLES SENF, Director of the Canal.

March 3.

City Gazette, March 7, 1797

The size of the workforce was increased by artisans hired for specialized tasks. Senf needed bricklayers, stonecutters, carpenters, blacksmiths, and others with similar specialties. Such artisans were in short supply in South Carolina, a predominantly agricultural state. Although there is no specific reference in Senf's reports, he probably had to recruit many of these artisans from the North.

As on most American canal-construction projects, sickness was a major problem. Senf himself became ill.[114] Deaths were frequent. Senf reported, "From the year 1793 to the beginning of 1800 twenty-four white Persons died at the Canal by Fevers, of which Number were two Physicians, two Assistants, three Commissaries, two Master Carpenters, three Master Bricklayers, and two head Overseers. The rest were Journeymen Tradesmen and Overseers."[115] These statistics were just for the white men. Fatalities among black slaves remained unreported but were probably much, much higher. According to Senf, the normal sickness rate was four per hundred. During the sickly season—late summer and early autumn—when warm weather increased the spread of disease, this number was probably far higher. At one point, probably in 1799, artisans or tradesmen were refusing to work on the Santee Canal during the sickly summer months: "The last year no Tradesmen of any Abilities would engage to remain on the Canal in the Summer Season, when the most Work could be done."[116]

Because of their high concentrations of workers in close, primitive quarters without adequate sanitation, canal workers were extremely susceptible to diseases, particularly water-related diseases and especially typhoid fever.[117] Although typhoid fever[118] has been almost nonexistent in the United States for the past one hundred years,[119] it was very prevalent during the eighteenth and nineteenth centuries. The workers were also susceptible to yellow fever, which is spread by mosquitoes.

The process of building a canal followed a logical pattern, to which Senf generally conformed in the construction of the Santee Canal. The first task was to build the houses, huts, and kitchens needed by the overseers and workers. In addition warehouses had to be erected for the company's supplies, as well as shops such as blacksmith shops. The next step was to clear the line of the canal and remove the stumps. Usually this involved clearing the land from fifty-five to seventy-five feet on either side of the canal's center line. Because the Santee Canal was a summit level canal, reservoir areas also had to be cleared before excavation could begin. The Great Reservoir at Kirk's Swamp, for example, involved clearing fifty acres of land prior to excavation.

Although some limestone had to be excavated, the soil that Senf was dealing with was generally soft earth. Nonetheless excavation, embankment, shaping, hauling, and puddling[120] a twenty-two-mile-long canal was intensive backbreaking work.

In the eighteenth century building-material suppliers did not exist. Senf had to find his own. Cypress timber for the lock gates had to be felled, pit sawn, and fabricated. Bricks for the locks had to be molded and burned on-site. Stone for the hollow quoins for the locks had to be cut and transported to the canal. Lime for the mortar used by the masons had to be burned, from shells or limestone, mostly found in the line of the canal. All this was accomplished by Senf's workforce.

For the first several years construction proceeded unhindered. Then, in January 1796, the Great Flood of 1796 swept away much of South Carolina's bridges, warehouses, and farms near the rivers. John Drayton described this landmark flood:

> And, in January 1796, a similar one [to the flood of 1701] came down the same river [the Santee]; ever to be remembered by the mischief it effected. No bridge could withstand the fury of its torrent; rendered more impetuous by the weight of large trees and houses, which were borne down by its stream. A wooden bridge over Broad River, a few miles above Columbia; and another (the third of the kind, which Mr. Wade Hampton with great perseverance had caused to be erected near that place) about seven hundred feet long, over the Congaree river at Granby, upwards of forty feet high above the common level of the river, and many of whose piers were fastened by iron bolts into solid rock at the bottom of the river, were swept away in the general ruin. At Granby, the tobacco ware house was destroyed; together with one hundred fifty hogsheads of tobacco which were therein. The Camden tobacco ware house, on the banks of the Watteree river, met the same fate. Dwelling houses, corn houses, cattle, horses, and hogs, were carried down by the violence of the current; and vast beds of sand, were strewn over fertile tracts of swamp land, to their irreparable injury. The collected waters, of almost all the rivers in the upper country, at length, effected a junction at the confluence of the Watteree and Congaree rivers; pouring down their consolidated turbid stream, with destructive velocity; rising at the rate of three inches an hour, and continuing to rise for some days. At this time, the current in a great degree swept directly down the swamp, in a width in some places, more than five miles from the high pine lands on either side; undirected by the course of the river where it made a bend across the swamp; and only following it when the direction was with the stream. Much provision was destroyed; thousands of bushels of Indian corn, and many hundred barrels of rice. Some of the negroe houses, of the lower plantations on Santee, were torn up, and were carried by the

torrent entirely out to sea. Rice plantations, within a few miles of the ocean, and on the best pitch of tide, were overflowed for near a week; the water being from two to three feet above the rice field banks; during which time the ebbing of the tides, for the first days, was scarcely perceivable. This great flood, insinuated itself on one side between Cat Island and the main, and entered Georgetown Bay between Cat and Ford's Islands; on the other side, it flowed into Hell-hole Swamp; and from thence entered the different bays, which communicate with the eastern branch of Cooper River.[121]

No company records of the impact of the Great Flood of 1796 on the Santee Canal and its workers have been located. The impact, however, must have been great. Hellhole Swamp, which Drayton said was flooded, was immediately adjacent to the canal under construction. The flood inundated most of the lowcountry. An account from a planter in St. Luke's Parish, near the New River, suggests what probably happened at the Santee Canal:

> "I have been more unfortunate this year than ever I have been ſince I commenced planter. I have been overflowed with the waters of deſolation, and they have deſtroyed all that I had on two plantations; rice, corn, potatoes, cattle, in ſhort, as compleat devaſtation as can be conceived. I think the height of the water was not leſs than 12 feet perpendicular, and ſo ſudden that it precluded our taking any meaſures to ſave any thing. In the evening of Saturday week my fields were dry; on Sunday the day following, I took my negroes from the roofs of their houſes and chimneys, all except one unfortunate man, who was drowned before I got to their aſſiſtance. I was obliged to tranſport a large boat upwards of five miles by land and two by water before they got relief. In my exertions I met with ſeveral accidents and eſcaped with difficulty from a watery grave; however, the paſt cannot be recalled, and I muſt ſubmit with patience. I am ſorry, to add, that there are too many in my ſituation, and ſome worſe."

City Gazette, February 6, 1796

It probably took Senf and his workers months to recover from this flood. They would have had to rebuild worker's housing, shops, and warehouses and to clear the excavated canal of deposited flood-borne debris and silt.

By 1797 flood cleanup was completed. New officers and a new board of directors were voted in at the January 1797 stockholders meeting.[122] A review of the construction found that the canal was about 75 percent complete—only a section of from five to six miles still needed to be dug. Senf estimated that with an additional sixty laborers and seventeen tradesmen, he would be able to complete the excavation of the canal and finish the needed locks, aqueducts, and culverts by January 1798. The board of directors unanimously voted to give Senf the extra monies needed for more workforce. In this spirit of optimism, a Mr. Chestnut of Camden built a canal boat based on plans drawn by Senf. Fifty-one feet long and nine feet wide, the boat was capable of transporting nine hundred bushels of corn or two hundred barrels flour or thirty-five hogsheads of tobacco through the canal. It was estimated that such a boat could be constructed for half the expense and, using the new canal, perform twice as many trips to Charleston as a normal Santee boat traveling down the Santee River to its mouth and thence to Charleston.[123]

Work delays persisted. The canal was not completed until 1800 (see fig. 2.13). The company announced this milestone in May of that year:

> **INFORMATION** is hereby given to all whom it may concern, that the Santee Canal will be competent to pafs Boats of the following dimentions, viz.
> Extreme length, 54 feet.
> Extreme breadth, 9 feet.
> Draught of water, not exceeding 2 feet 10 inches.
> *By order of the Board.*
> Adam Gilchrift, prefident.
> *Atteft,*
> John M. Davis, Secretary.

City Gazette, May 1, 1800

Later Senf would be criticized for taking such a long time to complete the canal, but its construction period compares favorably to other canals built at the time in the United States. The Potomac Canal, perhaps not as large a construction project, was under construction for seventeen years (1785–1802).[124] The Conewago Canal in Pennsylvania, a much shorter canal with only three locks, was under construction for five years (1792–97).[125] Neither of these other canals had the additional challenges of a summit-level section.

With the completion of the Santee Canal, the company hired a superintendent, eight lockkeepers, two receivers, and a boom keeper.[126] The company ordered Senf to extend chains across both ends of the canal and said it planned to charge an entry fee of one dollar to each vessel that would pass the chain barriers.[127] The company also announced: "That if any person shall willfully or maliciously cut, break down, damage, or destroy, and bank or other work, to be erected or made for the purpose of the said navigation; such person shall be adjudged guilty of felony, and on conviction, suffer death without benefit of clergy."[128] This death penalty was probably the most severe penalty for damage to any canal in the United States. There is no evidence that it was ever carried out.[129]

On May 28, 1800, the *Times* of Charleston carried the news of the first boat to transit the new canal: "We are happy in being able to announce to the public, that Mr. William Buford, an enterprising citizen, who lives on the banks of Broad River, near Pinckney Court House, which is more than ninety miles above Granby, arrived in this city, through the Santee Canal, on Tuesday the 26th inst. [that is, May 1800] with his own boat, built on his own land, and loaded with his own crop, after having safely passed over all the falls and shoals that are between his plantation and Charleston." The notice continued:

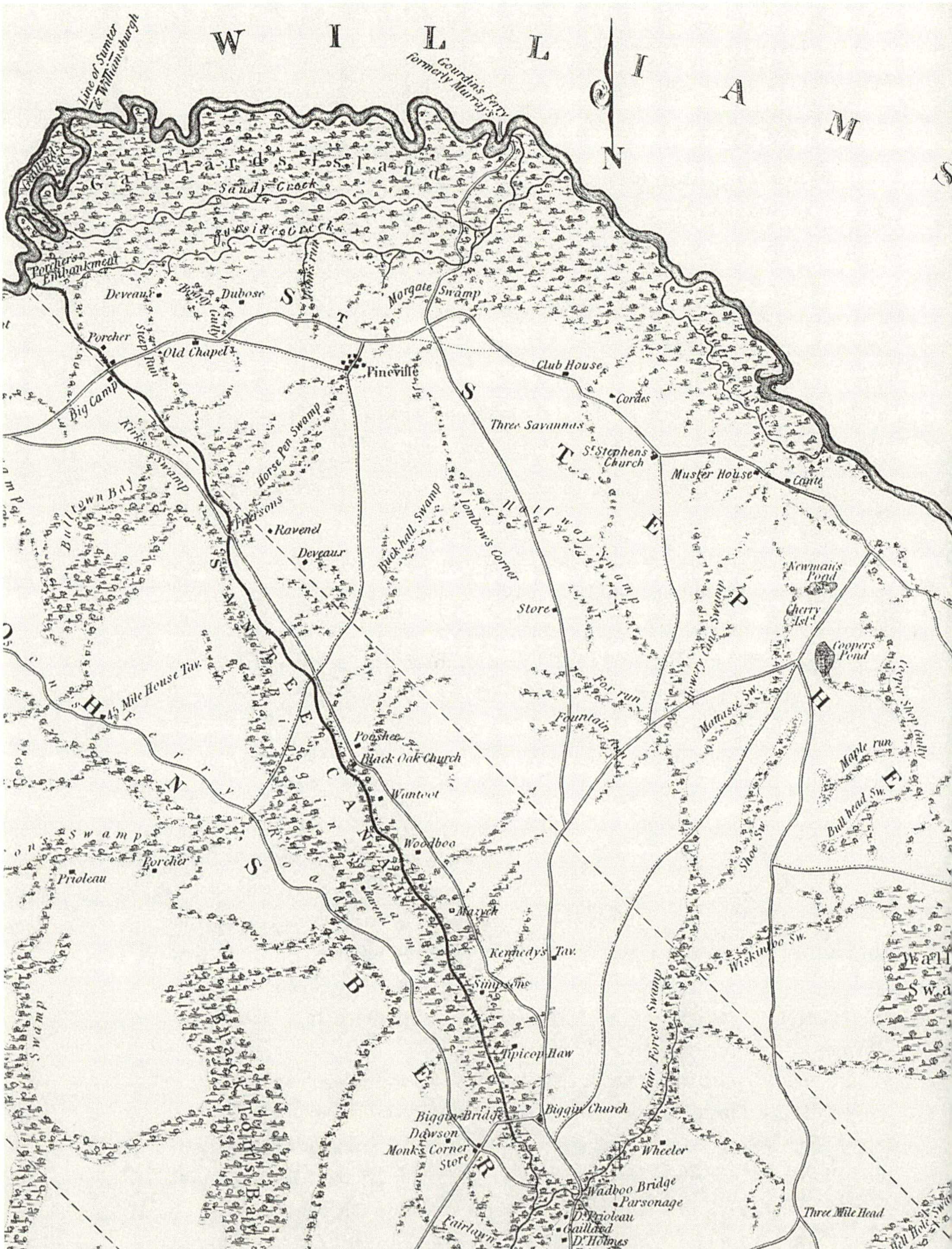

Fig. 2.13. The Santee Canal on Robert Mills's map "Charleston District, South Carolina, Surveyed by Charles Vignoles and Henry Ravenell, 1820, Improved for Robert Mills' Atlas, 1825."

This is the first enterprise of the kind that has taken place at so great a distance from the sea coast; and from its success, leads to the most pleasing anticipation of the immense advantages which are likely soon to result from the easy intercourse that may now be carried on, between Charleston and the interior country, by means of the Santee Canal.

When the obstructions of the navigation of the Broad and Catawba rivers, (which are far from being as numerous as have been represented) shall be removed, the superabounding productions of the upper country will flow into Charleston in such full tides, and with so much expedition and so little expense, as will lower our markets, and at the same time fill the pockets of our remote fellow-citizens. And what will be equally agreeable, the gentlemen who have persevered so successfully, and at such great expense to complete the canal will possess a property which while it contributes to the welfare of the state, will yield them an income that will amply repay all that they have expended in an undertaking which is unequalled in the new world![130]

The actual tolls established for the Santee Canal were much higher than the one dollar fee for passing the two chains. Senf recommended a flat toll of five pounds (about twenty-one dollars) for all boats traversing the canal.[131] The company however adopted a series of tolls based primarily on the breadth of the boat passing the canal (see table 2.1).

The final cost of the Santee Canal has been disputed but the most accurate figure seems to have been the one reported by the South Carolina Board of Public Works in 1823: $650,667.14[132]—almost three times Senf's original estimate of £55,620 (approximately $233,000). Senf has been criticized not only for the

TABLE 2.1. Santee Canal Company tolls, 1805

Type of vessel	*Extreme breadth*	*Cargo*	*Toll*
Boats or flats	not passing through entire canal		$2/lock; $3 double lock
Boats or flats	under 6½ feet		$10
Boats or flats	6½ feet–7½ feet		$15
Boats or flats	7½–8½ feet		$20
Boats	8½–9½ feet		$25
Flats	8½ feet–9½ feet	with fewer than 70 bales of cotton	$30
Flats	8½–9½ feet	with more than 70 bales of cotton	$35
Flats	8½ feet–9½ feet; having paid $30 or $35 shall be passed back		$20

Source: Santee Canal Company, *Report of the Board of Directors* (1805), 1–2.

amount of time it took to build the Santee Canal but also for this cost overrun, particularly by Porcher, who blamed the cost overrun on Senf's decision to supervise the workforce instead of using contractors: "It [the Santee Canal] was nearly four times that amount [Senf's estimate]. In constructing a work of this kind obviously the best and cheapest way of proceeding is to divide it into sections and let them out to contractors, who shall work under the general supervision of the engineer, and whose work shall not be accepted and paid for until approved by him. Col. Senf undertook not only the general supervision, but the immediate superintendence of the laborers, and in large measure assumed the functions of an overseer. Hence, of course, a great waste of time as well as loss of money."[133]

Porcher estimated the final cost of the Santee Canal to have been $40,000 per mile, or a total of $880,000. But Porcher included in this construction cost the amount of estimated unpaid dividends.[134] The actual per-mile cost was approximately $30,000, which compares favorably to other canals constructed at the time, such as the Conewago Canal in Pennsylvania, constructed for more than $100,000 per mile.[135] As with the criticism for the time it took to build the canal, blaming the cost of the Santee Canal on Senf may be somewhat unfair. Factors beyond his control drove up the cost. It was impossible for Senf to have anticipated the doubling of salaries during the last decade of the eighteenth century. In addition cost overruns were endemic to eighteenth- and nineteenth-century canal engineering. There was very little experience in canal building on which to base accurate estimates. The canal engineer was always under significant pressure to "lowball" canal construction estimates so as not to discourage potential investors. It was a rare early American canal that was constructed for the original estimate.

Though Porcher criticized Senf for directly supervising the construction of the Santee Canal rather than use contractors, at that time all American canals were being constructed in the same manner. It was not until twenty years later, when Benjamin Wright introduced the method of contracting on the Erie Canal, that contracting for construction work became the usual practice of canal construction in the United States. Senf was aware that contracting for the work was an alternative to direct superintendence. In a report to the Santee Canal Company's president and board of directors, Senf defended his decision not to use the contracting method:

> To have the whole or part of a Canal executed by Contract, is usual and frequently practiced in other Countries, but in the present situation, it is not advisable to adopt the same Mode. Here, where a Work can be done altogether by Contract, it is certainly best to have no distinction among Labourers. For instance, suppose, an Allowance was made, so much per Cubic Yard to dig and remove the Ground to a proper distance, by Negro Labourers, the making and forming the Dams could not be trusted to them, therefore, another party, sufficiently instructed, would have to do this Business. The Overseers, who overlook those Negroes, who work by Contract, will be

more apt to watch the Interest of their Employees and not that of the Company. Should the Ground change for the worse, or be obliged to dig a few feet deeper, where they could not do so much, as expected, an additional Allowance would be the immediately the demand, and as to Complaints and Disputes there would be no end. It is therefore my humble Opinion, that the most Certain and the most profitable Arrangement the Board can make, is to put all labourers on like Terms; and allow to every twenty negroes One Overseer.[136]

Operation of the Santee Canal

Once opened, the Santee Canal became a conduit for cotton shipped from the South Carolina upcountry to Charleston. Canal boats typically carried 70 to 120 bales of cotton, each weighing 500 pounds with dimensions of almost six feet long and several feet wide (see table 2.2).

Striking in this table is the high cost of tolls charged in 1823. At the time a skilled tradesman earned one dollar per day, a typical toll was eighteen to twenty times the daily pay of a skilled worker. Also of interest in this table is the large crews that were used to bring these boats through the Santee Canal. The availability of these large crews led the company to abandon their plans to use horses to tow the boats through the Santee Canal.[137] Instead the canal boats were pulled by their crews, almost always slaves (see fig. 2.14).

Fig. 2.14. "A Cotton Box at Black Oak Lock" on the Santee Canal, based on a drawing by Andrew Gibbes. Courtesy of the Charleston Museum Library

Table 2.2. Boats used on the Santee Canal

Boats	*Length*	*Breadth*	*Draft empty*	*Draft loaded*	*Bales cotton*	*Toll*	*Men*
1	56 ft.	9½ ft.	1½ to 2 ft.	3 ft.	80 to 120	$20	5 to 6
2	56 ft.	8½ ft.	1 ft.	2 ft.	80	18	4 to 5
3	56 ft.	7½ ft.	8 to 10 in.	1½ to 2 ft.	70	14	3 to 4
4	56 ft.	6½ ft.	8 to 10 in.	14 to 17 in.	50	10	3
Flats	54 ft.	9½ ft.	8 in.	2 ft.	120 to 130	60	5

Source: *Report of the Superintendent of Public Works for the Year 1823.*

Despite the high tolls charged, the canal company made no more than a modest profit. The company's annual expenses were approximately $7,000 per year. Against these annual expenses they earned a slight surplus as show in the following list of gross receipts from tolls for selected years:

1814	$21,249
1815	$18,459
1821	$15,180
1822	$11,031
1823 (to July 1)	$13,635

For the stockholders the Santee Canal was not a great success. The company was only able to issue a dividend in 1814 ($12,359) and again in 1815 ($13,086).[138]

In 1808 Secretary of the Treasury Albert Gallatin reported the reasons for the financial difficulties of the Santee Canal: "The want of success in this undertaking, which though completed is very unprofitable, may be ascribed to several causes. The expense, compared with the work, is much greater than might have been expected, and probably, than was necessary. The locks are too small for large boats, which are therefore obliged to pursue the former route down the Santee, and by sea to Charleston; and want of water is alleged as a sufficient reason for the size of locks. But a canal in that situation cannot in America be profitable, unless the navigation of the main river with which it communicates is rendered safe and permanent; and whenever that of the Santee itself shall have been improved the utility and profits of the canal will be considerably increased."[139] Such improvement of the Santee River did not begin for another two decades following the completion of the Santee Canal. This work was undertaken by the State of South Carolina.

Yet, although the canal did not generate expected dividends, it was not a losing proposition. As the list of tolls above indicates, income for the canal almost always exceeded its approximately seven thousand dollars of yearly expenses, and therefore the canal was self-supporting.[140]

Although the Santee Canal did not meet its investors' expectations of dividends, it did provide other economic benefits. For planters, many of whom had invested in Santee Canal Company stock, the company had hired their slaves at a time when agriculture was in decline and many of their slaves were idle. Most of the $650,667.14 spent on construction of the canal went directly to the planters for their slaves. The planters also profited by the proximity of their plantations to the canal. Advertisements of the time clearly indicate that a plantation on or close to the Santee Canal brought a higher price than a plantations not so situated. Charleston merchants, many of whom also invested in the Santee Canal Company, also benefited from the canal, which was completed just as the cotton boom hit the state. The canal greatly lowered the cost for transporting cotton from the upcountry to Charleston and eliminated the risk of losing a shipment in the ocean voyage otherwise required from the mouth of the North Santee or South Santee down to Charleston. The canal also reduced Charleston merchants' costs for shipping commodities such as salt and manufactured products to the upcounty.

Toll revenues dropped during the droughts of 1816 and 1818–19, when the canal was dry. As early as 1809, the directors of the company contemplated adding steam engines to pump water to the summit level, particularly during droughts.[141] Pumps were added later, but they proved less than satisfactory. Despite the problems arising from a lack of adequate water, the Santee Canal had years when as many as 750 boats passed through it.[142]

The Santee Canal was neither a complete success nor a complete failure; its financial performance was similar to those of other American canals built before the War of 1812. None of these canals was a huge financial success. *The Department of the Army Report of 1881* fairly accurately characterized the Santee Canal: "The canal was opened to commerce in the year 1800 and was operated with more or less success down to the year 1840."[143] However, the Santee Canal was to become an important link in the ambitious statewide river-navigation system initiated by the State of South Carolina after the War of 1812.

3 ~ The Catawba Company, the U.S. Military Establishment, and Early South Carolina Canals and Waterways, 1787–1812

> The companies, which are at present incorporated by the legislature of the state, for opening its inland navigation, are "The company for the inland navigation from Santee to Cooper River," "The company for improving the navigation of Edisto and Ashley rivers; and making a communication by a canal, and locks, from one to the other of the said rivers." And "The company for opening the navigation of the Catawba and Wateree rivers." Besides these, many are the legislative appointments of commissioners, throughout the state, for opening the navigation of rivers and water courses. So that, at this time, there is scarcely a river in the upper country, whose navigation is not more or less in a state of improvement.
>
> John Drayton, *A View of South-Carolina* (1802)

BEFORE THE WAR OF 1812, when South Carolina began to build canals and improve its waterways, the state either chartered private companies or used the levy to clear South Carolina rivers for navigation. In addition to the most successful of the chartered private companies, the Santee Canal Company, there were three others: the Catawba Company (chartered in 1787 for opening the Wateree and Catawba rivers), the Edisto Company (chartered in 1787 for opening a canal between the Ashley and Edisto rivers and for clearing the Edisto River), and the Sampit and St. James Santee Canal Company (chartered in 1809 for building a canal between the Sampit River west of Georgetown to the North Santee River). All these companies had difficulty raising financing in capital-short South Carolina.

In addition to these companies, the legislature authorized commissions to clear the state's rivers, giving them the power to levy. Typically the levy required all men from age six to sixty living within six to ten miles of the river to spend six days a year clearing the river of obstructions. The levy was very unpopular and was later limited to slaves. Compared to contractors used by private companies, the levy was also not very effective and by the end of the War of 1812, the levy was no longer used to clear waterways.

Most of these early efforts at canal construction and waterway clearance were not successful, but the experience gained through these projects helped to define South Carolina's massive canal-construction and waterway-improvement program after the War of 1812.

The Catawba Company and the U.S. Military Establishment

The Catawba Company, officially called "The Company for opening the Navigation of the Catawba and Wateree Rivers," was chartered on March 27, 1787, to improve the Catawba and Wateree and to build a canal to bypass the rapids near Rocky Mount,[1] thus connecting the two rivers. The Wateree flows into the Congaree to form the Santee River, and further downstream was the Santee Canal. If these waterways could be made navigable, commerce could flow to Charleston from as far away as North Carolina.

Very early the construction of the Catawba navigation became intertwined with the development of a new armory and arsenal the U.S. government was planning to build along the Catawba River. The armory needed a canal for transport and water power, and the canal company needed federal government assistance in funding the new canal. In 1798 President George Washington selected Rocky Mount on the Catawba River as the site of a third national armory and arsenal, writing: "It is conceived that three principal permanent stations will suffice and that these ought to be at Springfield and Harpers Ferry which are already chosen, and the vicinity of Rocky Mount on the Wateree in South Carolina. . . . It is expected that a canal will ere long effect a good navigation between the Wateree & the Catawba; which whenever it shall happen will render the vicinity of Rocky Mount extremely convenient to the supply of North Carolina by inland navigation."[2]

In 1802 Secretary of War Henry Dearborn[3] wrote to General Thomas Sumter,[4] at that time a U.S. senator, asking him to inquire about the possibility of purchasing land on the Catawba River above Rocky Mount for use as an armory and arsenal for the army: "Can you inform me whether a cession of the land at Rocky Mountain, has actually been made by the State, & whether a purchase is to be made by the U. States, if so, on what terms can it be had, say for twenty acres, so as to include the best Scite for the public buildings, and good water privilege?"[5] By June 1802 Dearborn had informed Sumter, "It is the intention of the President of the U. States to commence the establishment of a Magazine and Armoury near Rocky Mountain on the Catawba River."[6]

The site was to be relatively small—twenty to thirty acres, perhaps as many as fifty acres if the price was right.[7] Dearborn asked Sumter to select the necessary

Fig. 3.1. "View of the Great Falls of the Catawba." From Benson J. Lossing, *Pictorial Fieldbook of the American Revolution* (1859)

Fig. 3.2. "View at Rocky Mount." From Benson J. Lossing, *Pictorial Fieldbook of the American Revolution* (1859)

site and purchase it with the assistance of John Christian Senf,[8] former chief engineer of the Santee Canal and state engineer of South Carolina. Dearborn also wrote to Sumter, "In examining the Scite it is desirable that you and Colonel Senf should decide on the particular Spots for the principle buildings—some of which we shall commence immediately after the purchase shall be completed. The first building will be a moderate sized house for the Superintendant, and barracks for the workmen, which may be occupied ultimately by the armourors, a suitable Magazine for Powder, a Magazine for Arms and a Barrack for a small Guard, which will be detached from Charleston as soon as any public property shall be deposited there."[9]

This armory site was initially called the U.S. Military Establishment at Rocky Mount and later became known as Mount Dearborn.[10] The key to developing Mount Dearborn was the construction of a bypass canal around the rapids in the Catawba River above Rocky Mount, not only giving access to the site but also providing water power for the armory. Without this canal reaching Mount Dearborn would be difficult (see color plate 6 and figs. 3.1 and 3.2).

On August 9, 1802, based on reports from Sumter and Senf, Dearborn wrote to Sumter authorizing the purchase of a tract 300 to 500 acres.[11] In 1803 the U.S. government bought 523 acres on the west bank of the Catawba River in Chester County, greater than the 300 to 500 acres Dearborn had authorized and far greater than the 25 to 30 acres he initially envisioned.[12] The land transaction was somewhat questionable. Sumter probably marked up the price of the 250 acres he received from Isham Moore, executor of the estate of William Moore. In addition Sumter may have sold the government land that was actually owned by the Catawba Company.[13] This later became one of the points of contention between the canal company and the army.

Dearborn had intended to have Senf supervise the construction of the buildings at the new site. In November 1802 he wrote to Senf, telling him that "Mr. [Eli] Whitney who I believe you have some knowledge of and in whose Mechanical talents I have great confidence, and, who has lately erected an improved set of works for manufacturing Small Arms, will probably write you by my request from Raleigh in North Carolina on the subject of a meeting at Rocky Mount for the purpose of consulting in the most proper scite for the proposed buildings for the Armoury etc., more especially for that part of the buildings which will require water machinery"[14] (see color plate 7).

Because they had limited funds the Catawba Company had very early emphasized the clearing of the lower Catawba and not the construction of a bypass canal around the Great Falls of the Catawba. With the selection of the site for the new armory on the west bank of the Catawba River just above Rocky Mount at the falls, the Catawba Company became involved in the construction of a bypass canal. The U.S. Government would partially fund the construction of that canal to aid its new armory.

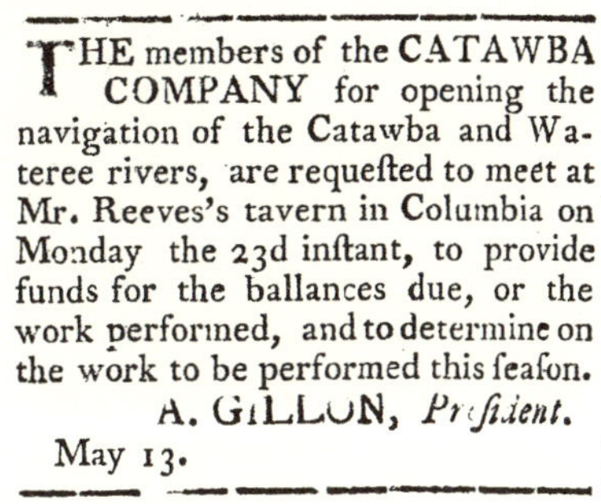
THE members of the CATAWBA COMPANY for opening the navigation of the Catawba and Wateree rivers, are requefted to meet at Mr. Reeves's tavern in Columbia on Monday the 23d inftant, to provide funds for the ballances due, or the work performed, and to determine on the work to be performed this feafon.

A. GILLON, *Prefident.*

May 13.

City Gazette, May 16, 1791

Work began on the armory in 1803. Because of Senf's experience in building the Santee Canal and because a major portion of the armory development was to be the construction of a canal, Dearborn offered Senf, who was also receiving pay as chief engineer of the Catawba Company, the superintendence of the construction at three dollars a day plus rations.[15] Dearborn wrote Senf in 1803:

> Mr. Whitney has made a report of the result of the examination made by him and yourself for a scite of a National Armoury and Magazine at Rocky-mount on the Catawba River, which is satisfactory.
>
> The first step now to be taken, is that the Canal company for opening the navigation of the Catawba river etc. shall agree to open the passage for the water, at the expense of said Company, from the river above the falls, into what is contemplated as the pond or reservoir of water for the water works of the Armoury etc. and which will form a part of the Canal and that the said Company agree to open said passage by the first day of October next, which agreement should be in writing subscribed by the members of the Company or their Agent and forwarded to this Office, and when the said Company shall have entered into such an agreement it will be proper (and not until then) to take measures for erecting a dam for our works, and for having a Saw Mill erected in what on the plan made by you and Mr. Whitney is contemplated as a waste way. . . .
>
> When the aforesaid agreements shall be completed for opening a passage for the water, if your other engagements will admit of your entering into the service of the United States for the purpose of superintending the erection of the buildings necessary for the establishment, it will be my wish to engage your services.[16]

Dearborn proposed a joint effort between the army and the canal company. In return for building the power canal, the Catawba Company would receive the benefits of the dam to be constructed on the Catawba River by the U.S. government.[17]

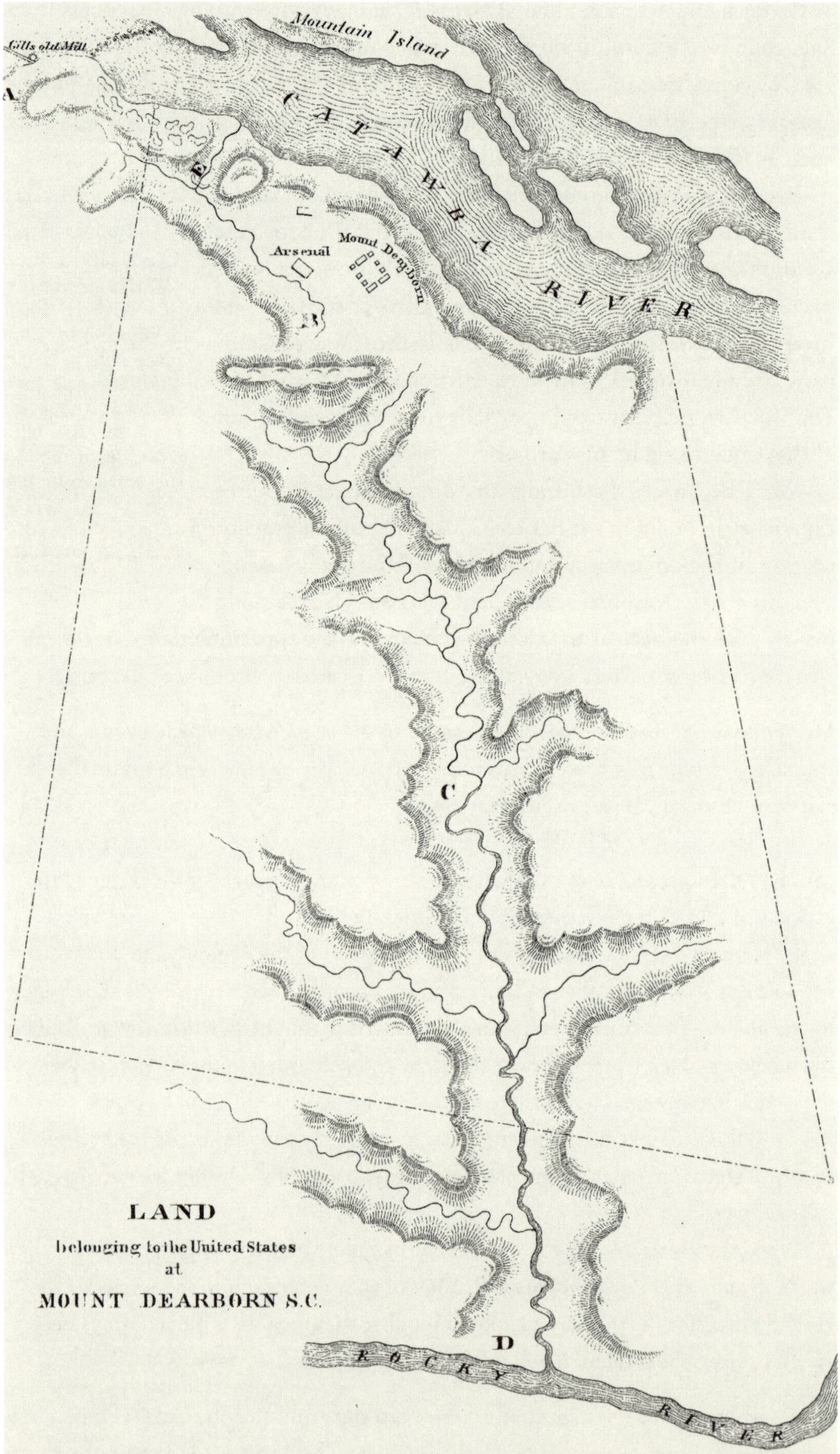

Fig. 3.3. Map of U.S. Army land at Mount Dearborn, prepared by the South Carolina Board of Public Works, ca. 1820. From *American State Papers, Military Affairs,* 3, no. 362. The route of the bypass canal around the Great Falls is marked from Gills Old Mill (upper left) through A, B, C, and D to Rocky River (bottom).

Both parties would profit. The army would gain transportation to the armory site and water power for its armory. The canal company would be able to use the canal for transportation around the falls in the Catawba River. The army had entertained similar cooperative arrangements with private companies at its other armories, such as the proposed but not implemented Shenandoah Canal connection with the power canal to have been built by James Brindley alongside the Harper's Ferry armory.[18]

Senf referred Dearborn's proposal on the bypass canal to Judge John F. Grimké,[19] president of the Catawba Company, in March 1803 and advised him, "It is no great object in regard to the expence; Five hundred dollars will effect the purpose: Twenty persons will do this work in Six weeks."[20] Five hundred dollars was a surprisingly low estimate for the cost of this canal. Having been chief engineer of the Santee Canal and now the most experienced canal engineer in the South, Senf should have known that the Rocky Mount Canal would be far more expensive. Nevertheless an agreement was made, and the Catawba Company agreed to provide five hundred dollars for the work, which, it was thought, could be finished by October 1, 1803.[21] But the canal was not finished by that date, nor could it be constructed for the estimated five hundred dollars. This underestimate later became a source of contention between the canal company and the army (see figs. 3.3 and 3.4).

Meanwhile work began on the buildings needed for the new armory. On May 6, 1803, Dearborn wrote to Grimké about Senf, "There will be no objection on the part of the United States to his [Senf's] dividing his time between the public works and the Canal, if the respective works should be in the vicinity of each other."[22] On May 14, 1803, Dearborn wrote to Senf, passing on this information and confirming his appointment as superintendent for the construction of the buildings at Rocky

Fig. 3.4. Sketch of the U.S. Military Establishment at Mount Dearborn, by Robert Mills, ca. 1825. Courtesy of Tulane University. The view is to the north along the route of the proposed bypass canal.

Mount.[23] By November 1803 Dearborn had sent to Senf, via Eli Whitney, detailed specifications for the buildings to be erected.[24]

For their part the Catawba Company began to recruit an engineer to assist Senf in building the canal. William Loughton Smith[25] wrote to architect-engineer Benjamin Latrobe in Philadelphia on March 21, 1804, and John Vaughan[26] wrote to Latrobe on March 14—both asking Latrobe to recommend to them an engineer for the Rocky Mount Canal.[27] On March 18 Latrobe enthusiastically recommended Frederick Graff[28] to Vaughan: "If he is [willing to work for the Catawba Company], the Company will obtain the assistance of *the best Superintendant, I ever knew,* of a man upon whose integrity and knowledge I had calculated largely in the superintendance of the Chesapeake and Delaware Canal, and whom I would not spare did I not conceive his own interests likely to be promoted by having a separate command."[29] Latrobe added that he would immediately inquire if Graff's services could be obtained, under what terms and whether he could immediately proceed to Charleston with, "two or three practical and Laboring blowers of rock."[30]

On March 21 Latrobe reported to Vaughan and Smith that Graff had accepted the position,[31] and by the summer of 1804, Grimké was reporting to Dearborn that good progress was being made in the construction of the canal.[32] Graff's appointment to the Catawba Company did not work out. Approximately a year after he arrived in South Carolina, he resigned his position and returned to Philadelphia, leaving behind an uncompleted canal.[33] Grimké attributed Graff's departure to Graff's being "hypochondriacally inclined."[34]

Work also lagged on the buildings at the U.S. Military Establishment. In March of 1805, Dearborn decided to send Senf an assistant, Lieutenant Colonel Francis Mentges, to expedite the work.[35] Mentges arrived at Rocky Mount in April 1805 with the assignment to supervise the workmen constructing the buildings. However, Mentges died a few months later, on September 6, 1805.[36]

Dissatisfied with Senf's performance in building the armory and canal, Dearborn decided in 1806 to reassign the supervision of the construction to the Army Corps of Engineers.[37] This responsibility was assigned to Captain (later Major) Alexander Macomb.[38] Macomb took over the duties of superintendent for the Rocky Mount Armory from Senf in April 1806.[39] Senf's health deteriorated soon after, and he died on August 24, 1806, at age fifty-three.[40] Although Senf had not completed the needed canal, neither had any of the other engineers assigned the task. It was a very difficult task and was not completed for another twenty years.

Meanwhile work continued on the armory and arsenal. Dearborn provided Macomb with detailed instructions on how to build the magazine[41] and the other needed buildings.[42] Progress on the buildings at the new armory languished. Part of this delay was the lack of skilled workmen.[43] At one point Macomb augmented his workforce with escaped slaves from Georgia, who had been captured by the Indians of the region and delivered to him.[44]

At this time Macomb, in an obsequious gesture to his supervisor, gave the name "Mount Dearborn" to the U.S. Military Establishment along the Catawba River: "The U.S. Establishment on the Catawba has no particular name and is too far from Rocky Mount to be called by the same name. I beg you will permit me to honor it with the name of Mount Dearborn a name that will be extremely agreeable to the Inhabitants of the vicinity & very grateful to those employed at the works," Macomb wrote to Dearborn on February 27, 1807.[45] There is no record of what Dearborn thought about having the arsenal and armory named after him, but the name stuck.

The relationship between the army and the Catawba Company became extremely contentious over two issues: title to the land the army occupied and failure to complete the canal. It appeared that approximately two hundred acres of the land Sumter sold to the government was actually company property, and the company demanded that the government relocate from that land. Meanwhile the army demanded that the company live up to its agreement and finish the canal. Lawsuits were threatened: "I am commanded by the Honorable the Secretary of War to inform the Catawba Company, that neither the reasons offered to Colonel Williams, nor any threats of suits to be commenced will induce a removal of this Establishment to relieve the Company from their engagements to open a passage for the water through the ridge above our Dam," wrote Macomb to the company on June 9, 1807.[46]

Grimké answered for the company on July 6. First, he wrote, "The two subjects are totally unconnected."[47] With respect to the controversy over the canal, Grimké argued that the company had fulfilled its obligations, and to buttress his position he quoted from a letter Senf sent the company in 1803: "I am confident that you & the Catawba Company have the [illegible] Country's interest so much at heart, that you will not lose any time to give a favorable answer, and furnish me with the means to open the door of navigation at D. It is no great object in regard to the expence; Five hundred dollars will effect the purpose; Twenty persons will do this work in Six weeks."[48]

Grimké then quoted from a letter he sent to Dearborn on April 1, 1803: "They (the Company) are pleased to find a beginning was likely to be made in this usefull work, & they unanimously voted the sum of $500, for the purpose of opening the channel mentioned in your paragraph."[49]

Grimké continued:

> All that we can do, we have done, by first [illegible] making the means required, by the calculation of Col. Senf for the opening the said channel, <u>to him as your agent</u> [emphasis in the original], for whose punctuality we do not think you ought to

expect us to be accountable. On the 19 June 1804 I again wrote the Secretary of War as follows—I have now to observe to you, that we employed a number of hands for the purpose of digging through the Barrier at Gile's Falls, but they were impeded by the means of some misunderstanding with Giles, of which I suppose Col Senf has informed you.... Accordingly Col. Senf proceeded to some of the work, & expended the 500 dollars which he calculated it would cost, & which we paid to him for the purpose of opening the Barrier. The Catawba Company have done what they promised to do.[50]

There the matter remained. Nothing had gone well at Mount Dearborn, for either the army or the canal company. The canal had not been completed, and the company lacked funds to complete it. Without the canal the army could not power its machinery or even transport machinery to the site. As Major Macomb reported in 1809: "The great misfortune *at present* is the difficulty of communication, both by land and water, for it is placed in the midst of all the obstructions of the Catawba, so that you can neither go up or down the River, and consequently until they are avoided by the canal the establishment will be deprived of easy access to the iron works of the upper country as well as the free communications with the lower country."[51]

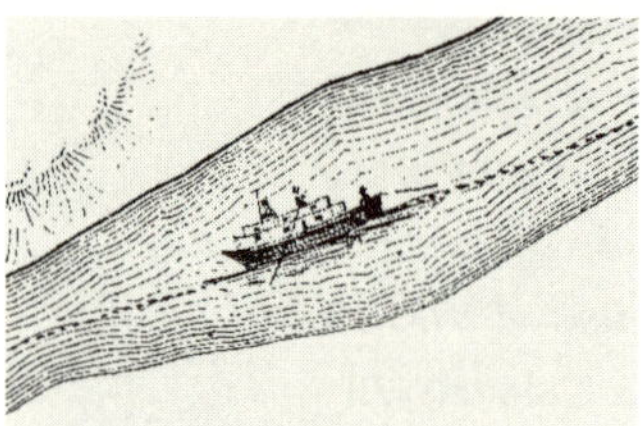

Fig. 3.5. Detail of a river craft on the Catawba River, from "A Survey of the Establishment of the United States at Mount Dearborn" (1809), by Alexander Macomb. From *American State Papers, Military Affairs,* 3, no. 362. One of the boat's crew appears to be poling while the second is steering with a long steering oar.

Fig. 3.6. Detail of a second river craft on the Catawba River, from "A Survey of the Establishment of the United States at Mount Dearborn" (1809), by Alexander Macomb. From *American State Papers, Military Affairs,* 3, no. 362. Two crew members appear to be rowing while the third is handling the long steering oar.

Attempting to secure financial assistance to complete the canal from the federal government, the company on September 22, 1808, submitted a memorial to Congress to subscribe to its stock.[52] On January 26, 1809, Congress turned it down.[53]

This situation continued for some time. The canal company continued to contest the title to the land of part of the site. With the canal uncompleted there was no water power for the armory. The buildings never served as armory or arsenal; they were garrisoned during the War of 1812 and abandoned sometime thereafter.

The concept of a river navigation on the Catawba River using bypass canals at the Great Falls was resurrected after the War of 1812 as part of South Carolina's canal-construction and waterway-improvement program. There were plans to link this river navigation to a Catawba River navigation in North Carolina, but that portion was never built.[54] The South Carolina portion was built and used. The bypass canals around the falls at Rocky Mount were finally completed by 1828 (see figs. 3.5 and 3.6).

Other Early South Carolina Canal and Waterway Projects

In 1753 the South Carolina General Assembly passed an act titled "An Act for Appointing and Impowering Commissioners to make the Wateree River Navigable."[55] The appointed commissioners—Samuel Wyly, Robert Millhouse, Jared Neilson, Matthew Neilson, and James M'Goit[56]—were authorized "to take away the rafts and other obstructions in the said river, or to make the said river navigable, by cutting a water-passage in any other manner as they shall judge most expedient."[57] This work was to be accomplished through the use of the levy: "All the male inhabitants, from the age of sixteen to sixty years, living and residing on either side of the said river, within ten miles of the same, and not more than seventy miles above the said rafts, or either of them, shall be, and they are hereby, obliged and required to work in removing the said rafts, or making the said river navigable, in any other manner as the said commissioners, or a majority of them, shall judge most expedient."[58]

Men refusing to participate in this work were subject to fines.[59] The levy would last for a maximum of six days per year.[60] This effort had only limited success. In 1784 broader legislation was passed to clear Wall's Cut, and the Edisto, Wateree, Great Pee Dee, and Little Pee Dee rivers through the use of the levy.[61] Again in 1791 legislation was passed to open not only the Wateree but also the Great Pee Dee, the Congaree, the Broad, the Savannah, and other rivers through the use of the levy.[62] Commissioners were appointed for the various rivers. This and other levies were extremely unpopular and came to be interpreted as applying only to slaves. Seen as an inexpensive expedient, the levy was never very effective in opening South Carolina's waterways to navigation.

There were other attempted canals in the early days of South Carolina. An important one was the proposed Edisto-Ashley River Canal intended to link the Edisto River with the Ashley River, thus creating a route to Charleston. Surveys of a possible route were undertaken as early as 1786. According to the *Report of the Civil and Military Engineer for 1818:* "Several plans have been suggested to effect this object. As early as the year 1785, the attention of the legislature was directed to this subject, and a survey was made by their order in 1786. The course recommended in that report, ran through the Cypress Swamp, where the distance does not exceed eight miles. Another plan was presented at the same time, which brings the canal out lower down the Ashley. The distance by this route is ten miles and three quarters."[63]

At a January 2, 1787, meeting, a company was organized "for cleaning and making navigable Edisto River, and the Forks thereof—for opening a communication between Edisto and Ashley Rivers, by a Canal and Locks, and for clearing the latter."[64] They proposed that the new company be funded by the sale of one thousand

At a meeting of a number of Gentlemen, holden at the State-Houſe, on the 2d day of January, 1787,

The following Plan,

For cleaning and making navigable Ediſto River, and the Forks thereof—for opening a communication between Ediſto and Aſhley Rivers, by a a Canal and Locks, and for clearing the latter, was agreed to.

I. THAT it be undertaken by a company, conſiſting of one thouſand ſhares.

II. That ſubſcription papers, conſiſting of one hundred ſhares each, be ſet on foot to have filled up as ſoon as poſſible.

III. That Col. Eveleigh, Col. Hammond, and James Lincoln, Eſq. be a committee to put in execution the foregoing propoſitions, and alſo to prepare a repreſentation and petition in behalf of the company, to be preſented to the general aſſembly at their preſent meeting, praying to be incorporated with proper powers for carrying the ſaid plan into effect, and to have an adequate toll granted to them.

IV. That a meeting of the ſubſcribers be holden on the 13th inſtant, in Charleſton, to conſider of, and determine upon the ſaid repreſentation and petition; at which meeting, each ſubſcriber ſhall have a right to vote for every ſhare taken by him; and abſentees may vote by proxy appointed by writing, for each ſhare which they hold; and a majority of the votes then preſent (given according to ſhares as aforeſaid) ſhall be concluſive and obligatory on all the ſubſcribers.

Fig. 3.7. Announcement of the Edisto-Ashley River Canal Company, 1797. From the Joel R. Poinsett Papers, 12/22/1, South Carolina Historical Society. Courtesy of the South Carolina Historical Society

shares. A committee of three—Colonel Nicholas Eveleigh, Colonel Leroy Hammond, and James Lincoln—was named to prepare a representation and petition to the General Assembly.[65]

On March 27, 1787, the General Assembly passed legislation establishing a company "for clearing and improving the navigation of the Edisto and Ashley Rivers, and for forming a communication by a Canal and Locks between the former and the latter."[66] This company was usually referred to as the Edisto Canal Company (see fig. 3.7).

Both of the two routes surveyed in 1786 between the Edisto and Ashley would have exposed the proposed canal to the floods of Cypress Swamp. The Edisto Canal Company identified a third route: "A proposal was afterwards made to open a canal from Hart's Cove on the Edisto, to Slann's bridge, a route not only exposed to freshets, but bringing the canal into Ashley, above the influence of the tide, and above some obstructions to the navigation of that river, which it would be both difficult and expensive to remove."[67] At a February 27, 1788, meeting, the company began to examine the various routes. However, it became apparent to the members present that construction of the canal was beyond their financial resources: "But it was the clear Opinion of the members present that nothing could be done for the present with respect to the [Edisto-Ashley River] Canal—And that it would be most Expedient for the Company to apply their Strength and their resources to Open & Clearing the Navigation of the [Edisto] river itself."[68]

On March 28, 1788, the company approved a newspaper advertisement asking for proposals to clear the Edisto River for navigation.[69] Several respondents proposed to clear the river for five hundred pounds.[70] No action was taken by the company.

By 1809 the issue of a canal between the Ashley and Edisto rivers had again come to the fore. In that year a John Diamonde proposed to survey the country between the two rivers for three hundred dollars.[71] Thomas Dillan, William McIntosh, and others also made proposals to survey that area.[72] The Edisto Canal Company employed surveyors Eve and Trotti to do the job and to recommend a practical route for the canal:

> Messrs. Eve & Trotti were employed by the Edisto Company in the year 1809, to survey the country between these rivers. The result of their examination has been repeatedly before the public. The sketch No. 8, will point out the route which, after examining all the above plans, and inspecting the ground, appears to me the least objectionable. Hart's Cove, situated between Givham's and Parker's ferries, is the most eligible site to take the canal out of the Edisto river, uniting the advantages of an accessible entrance and a safe and commodious harbour for boats.
>
> The two first miles present some difficulties, and there will be deep cutting for half a mile. The remainder of the route passes through old rice fields to within one mile of Ashley river, where the country is wooded. The canal will enter the Ashley river below all obstructions to its navigation. The length of this line will be eleven miles and three quarters.[73]

The company never built the Edisto-Ashley River Canal. The idea of building such a canal was reexamined in the 1820s as part of South Carolina's canal-construction and waterway-improvement program, but no work was undertaken.

Coastal navigation was also addressed during this early period. On December 19, 1809, the state legislature chartered the "Company for the Inland Navigation, from Sampit into Santee, and from Santee into Cooper or Wando River"—the Sampit and St. James Santee Canal Company.[74] The canal would have left the Sampit River west of Georgetown and proceeded south to the Santee River, thus providing an inland navigation from Winyah Bay to the Santee River. This company never constructed the authorized canal. In 1816 the Winyaw and Wando Company was incorporated to build a canal between the vicinity of Georgetown and Winyah Bay to the Santee Canal and from there to the Cooper River or Wando River, thus permitting navigation to Charleston.[75]

Prior to the War of 1812, the Broad River was used as a navigable waterway. At the northern end of Broad River, Lockhart's Shoals with a forty-seven-foot fall of the river, blocked navigation. In 1813 the state legislature passed legislation to construct a portage there. The act appointed and empowered commissioners to construct the portage on the western side of the river—where Lockhart's Canal was constructed by the state a decade later.[76]

These early efforts at canal construction and waterway improvement in South Carolina were largely unsuccessful. Either financing could not be secured or work mandated by the levy was ineffectual. Other canal projects initiated throughout the United States before the War of 1812 met similar fates. Either the canals were never completed, or, if completed, they were not financially successful. The situation changed dramatically after the conclusion of the War of 1812. Yet, although the prewar canals and waterways were unsuccessful, they helped to create a collective consensus on what waterways needed to be built and a body of experience on how to build them.

After 1814 the use of the levy for waterway clearance was abandoned in favor of hiring contractors for the work. This required money, but the War of 1812 had brought new wealth to the United States, money that could be marshaled to fund new waterways. The end of the War of 1812 coincided with the end of the Napoleonic Wars, after which European investors became willing and able to invest in American canal projects. New York, Pennsylvania, and many other states developed ambitious programs for internal development. South Carolina's internal-improvements program built numerous canals and opened more than two thousand miles of waterways in the state to navigation.

4 ~ Building the South Carolina Canal and River-Navigation System, 1817–1828

The utility of connecting all parts of the State with the Sea-Board—the benefit, pecuniary and political, of uniting in the bonds of commerce the people of the interior with the commercial metropolis, seem too evident to require illustration. The expense of effecting our plan of Internal Improvement, must be repaid in a very few years a hundred fold. It must, if accomplished, stimulate the enterprise of the citizen and exalt the character of the State. It must develope to the full all the capacities of its soil, and increase the variety and augment the quantity of its productions. There is no quarter of the State but what will participate in the benefits of the scheme; there are none of its lands, however remote from a market for their produce, but what will be enhanced in value. There is no portion of its cultivators, however distant from the mart of their annual supplies, but what will receive them at a reduced rate, in consequence of the diminished cost of transportation, and effect a corresponding saving in conveying the produce of their industry to the place of sale.

Editorial, *Southern Patriot,* December 5, 1820

THE WAR OF 1812 ENDED with the signing of the Treaty of Ghent on December 24, 1814. The war had been profitable for the United States. Individual Americans had gained greatly from supplying the warring European combatants. State treasuries, including South Carolina's, were flush with money. The treaty opened immense lands in the West for settlement, and Americans turned toward finding means to ship the agricultural produce of these western lands to the seaports along the coast. This meant improving transportation, specifically canals. New York State began work on the Erie and Champlain canals. Pennsylvania initiated the Pennsylvania Main Line Canal. Maryland began the Chesapeake and Ohio Canal. Virginia resurrected the James River Canal. And South Carolina launched an ambitious internal-improvements program to clear its rivers for navigation and to build bypass canals around the principal river obstacles.

In 1816 Governor David R. Williams[1] proposed such a program for South Carolina. The following year the new governor, Andrew Pickens Jr.,[2] suggested that the large surplus in the state treasury be applied to funding an ambitious internal-improvements program. In that same year the state legislature established the position of state civil and military engineer. John Wilson, a graduate of the University of Edinburgh and former major in the U.S. Army, was hired for that position, becoming the first supervisor of the state's canal-construction and waterway-improvement program.

In December 1818, the South Carolina legislature passed an omnibus improvement bill, which read in part:

Whereas, it is expedient to appropriate one million dollars for the Internal Improvements of the State.

I. Be it therefore enacted . . . That the faith of the State is hereby pledged to appropriate and set apart one million dollars for the Internal Improvement of the State, and that the sum of two hundred and fifty thousand dollars, annually, be, and the same is hereby appropriated for the term of four years, to be paid out of the annual balances in the treasury, and from the yearly dividend of the Bank of the State of South Carolina.

II. And be it further enacted . . . That the said sum of two hundred and fifty thousand dollars per annum, for the term of four years, shall be laid out and expended in improving the navigation of the rivers and other water courses of the State, in opening and constructing turnpikes, in cutting canals, and in such other works as will facilitate the transportation of the production of the soil to market, in such manner as may be prescribed by the legislature.[3]

This appropriation was also used to build new roads, but its primary emphasis was on creating river-navigation channels and, where needed, bypass canals around river obstacles. More than two thousand miles of river navigation were opened, allowing virtually all areas of the state to be reached. Not only did these improvements open up all of South Carolina, but—coupled with similar developments in North Carolina—they also opened commerce to the foot of the Blue Ridge through the use of the Broad River navigation,[4] the Catawba River navigation,[5] and the Great Pee Dee River navigation[6] (see color plate 1).

South Carolina's million-dollar internal-improvement plan was ambitious. One writer of the time described its expected benefits: "The improvement of our navigable rivers will afford facilities in the transportation of our surplus products, by which all classes of citizens in our State will be permanently benefited. The profits of agriculture will be increased, the resources of our commercial towns will be extended, industry will be stimulated to new and successful exertions. Our population will no longer be drained by emigration to other States, and the wealth and enterprise of our citizens, employed in supporting the liberal policy, which has been adopted, and in exploring new sources of public prosperity, will advance the character of our State to a rank which shall justify the hopes and exertions of an enlightened people."[7]

There was reason for such optimism. The price of cotton was high, and new lands were being developed for cotton production in the upcountry region. Success of the initiative seemed assured. For the times one million dollars for internal improvements was an immense sum of money, and an additional nine hundred thousand dollars was later appropriated to complete these projects.[8] Such extensive funding made South Carolina's program one of the largest of its kind in the United States at that time, particularly when it was compared on a per-capita basis with other states' programs.

The South Carolina internal-improvement program was first supervised by John Wilson. In 1819, however, the legislature established a Board of Public Works to replace the position of state civil and military engineer. The first board consisted of Wilson; Joel R. Poinsett, a representative to the General Assembly; former governor David Williams, then working on improving the Great Pee Dee River; William R. Davie; and Thomas Baker. In the next year the board underwent several personnel changes: Wilson resigned because of poor health; Williams declined to continue to serve and was replaced by Abram Blanding, who became commissioner of roads, rivers, and canals; Davie died;[9] and Poinsett was elected president of the board.[10] Further changes occurred in 1821 and 1822. Poinsett resigned as president in 1821 to assume his seat in Congress, and in November 1821 Nicholas Herbemont became president of the board. He was joined on the board by Robert G. Mills of Chester; architect and engineer Robert Mills of Baltimore; and in 1822 state senator William J. Myddleton.[11] The staff of the board in 1820 consisted of John Wilson, acting commissioner for the Department of Roads, Rivers and Canals; Thomas Baker, acting commissioner for the Department of Public Buildings; Thadeus Sobieski, engineer; John Couty, assistant engineer; and W. M. Gibbes, secretary of the board and treasurer of the contingent fund.[12] In 1822 the board was superseded by the superintendent of public works, and Abram Blanding was appointed to that office, a position he held for the duration of the internal-improvement program.[13]

At its first meeting, held on January 24, 1820, the Board of Public Works established its three divisions of work:

First. Opening and improving the navigation below the falls.
Second. Opening the navigation by canals, locks, and sluices above the falls.
Third. Constructing a road according to the Act of the Legislature, from Saluda mountain by Columbia to Charleston.[14]

These divisions of work reflected the direction provided by the legislature. In developing their plans for an internal-improvement program, the legislature intended to develop waterway system for steamboats, canal boats and other vessels that would use existing rivers for navigation as much as possible. Above and below the falls of the principal rivers—such as the Congaree, the Santee, and the Great Pee Dee—passages would be cleared through sandbars and obstacles would be removed. Bypass canals would be constructed around the major rapids and falls. These bypass canals were to be short, typically only a few miles long. While steamboats could be used below the falls, they could not fit through the canal locks. The canals would be equipped with lift locks capable of passing a canal boat seven feet wide and sixty-four feet long.

Where New York, Pennsylvania, and other states developed long still-water canals such as the Erie Canal and the Pennsylvania Canal, South Carolina's approach to the development of waterways as navigation systems was different. South Carolina used its rivers for navigation and built still-water bypass canals around major obstacles. Still-water canals were typically separate from existing rivers and were equipped with lift locks, usually of masonry. Extensively used in Europe, such canals were frequently recommended by European engineers for use in the United States. Because still-water canals required more excavation and support structures, still-water canals were more expensive than improved river navigations. It was thought that the State of South Carolina could save money by using its many rivers as navigable waterways and supplementing them with short still-water bypass canals. Wilson expressed this opinion in his 1818 report: "Before we treat of the method of removing these obstructions, it may be necessary to premise, that the state of this country does not authorise the expence, nor require the very costly works generally recommended by European Engineers. That a navigation through the mountainous districts ought to be confined to boats carrying forty to fifty bales of cotton, say sixty-four feet long, by seven feet wide, with at least eighteen inches draught of water at all seasons of the year."[15]

Using rivers as navigable waterways also meant that sluices—inclined ramps and channels in the bed of the river—would have to be created. The Board of Public Works intended to use sluices wherever possible: "In those parts of the river where the current is regular, and where the fall does not exceed seven feet in a mile, sluices may be made by low wing dams. In all cases where it is practicable the sluices should be formed near the bank of the river, so as to enable the boatmen to haul the boats, either by a tow line or windlass."[16]

The board agreed that lift locks, which were expensive to construct, should be used only at "some few places": "Whenever the fall is too great to admit of sluices, dams must be carried across the stream, so as to pen up the water from one shoal to another, and Locks so fixed as to be protected from the freshets. The Locks should be eighty feet long by ten wide; and the greatest lift ought not to exceed eight feet. In some few places it will be necessary to make a navigation round the falls by a lateral Canal, as at the falls of Rocky Mount on the Catawba river."[17]

This preference for river-navigation systems over still-water canals, was challenged by architect-engineer Robert Mills, who had recently arrived in South Carolina from Baltimore to become a member of the Board of Public Works. In 1821, shortly after his arrival in South Carolina, Mills published *Inland Navigation,* a pamphlet proposing a still-water canal that would not use rivers for navigation. This canal would be 110-miles long and would connect Columbia, in the heart of the state, with Charleston.[18] Mills proposed that the water from this canal would also provide additional water for the rice fields as well as water for the Santee Canal, at that time adversely impacted by the recent droughts of 1816 and 1818–1819. The water from this proposed canal would also provide water power for manufactories to be constructed between Columbia and Charleston. Mills estimated that $1,130,000 would be required to construct such a canal.

Mills's proposal for this still-water canal was widely circulated, and his ideas drew much interest. A bill was introduced in the legislature in December 1821 providing "for a survey from Granby to Charleston on the route delineated by Robert Mills for a canal, the preparation of a list of owners of the land through which the same would pass, and an estimate of the expense of the work."[19] The bill failed to pass, and that was the end of Mills's proposal.[20]

The official response to Mills's "Great Canal" came from Abram Blanding in the 1823 report from the superintendent of public works:

> It is a conceded fact, that a *river* navigation is always inferior to a *canal* navigation. And it is thence concluded by some, that the last should be substituted for the first. This conclusion, as applicable to South-Carolina, is however certainly erroneous, and should not be encouraged at the present day. I fear it will never be correct. Our country does not support a population, and I think never will, that can warrant opening a canal running between two navigable streams. To connect by a *canal* an extensive *river* navigation with our commercial capital, or to connect the upper with the lower part of a river, having good navigable water above, by a *canal,* round its falls, is all that can be expected from our local situation, the state of our population, or the best attention to our interests as an agricultural or commercial nation. We can have no *grand canal*—precisely for the reason, that our circumstances do not demand one.[21]

While at the Board of Public Works, Robert Mills was used primarily as an architect rather than as a canal engineer. He spent almost no time on waterway improvements or canal construction, with the exception of preliminary surveys to connect the Savannah and Broad rivers, the supervision of the construction of the Columbia Canal, and the certification of work completed on the Landsford Canal.[22] Instead he designed and supervised the construction of a series of courthouses and other buildings throughout South Carolina.

The first of the three work divisions identified by the Board of Public Works at their first meeting, "opening and improving the navigation below the falls," was intended to facilitate the steamboat travel on South Carolina's waterways, as well as navigation by canal boats and other vessels. There were five possible steamboat routes in South Carolina to channel commerce, primarily cotton, to Charleston: (1) from Granby,[23] below Columbia, via the Congaree and Santee rivers and then down the coast; (2) from Camden via the Wateree, and Santee rivers and then down the coast; (3) from Cheraw via the Great Pee Dee River and then down to Charleston, initially via the Atlantic Ocean and later through the Winyaw and Wando Canal; (4) from Savannah via Beaufort and Edisto Island; and (5) from Hamburg, opposite Augusta, Georgia, down the Savannah River.[24]

Funding for the initial four years of this program (1819–22) greatly favored opening the rivers of central South Carolina, such as the Congaree and Santee rivers, thus facilitating steamboat travel from Granby to Charleston and back and from Camden to Charleston and back. In later years (1823–28) funding favored other rivers, such as the Great Pee Dee and its tributaries, although the sum spent on these other rivers was substantially less than that spent on the Congaree and Santee rivers (see tables 4.1 and 4.2).

State engineer Wilson was one of the people who strongly believed that the initial work should be concentrated on the Congaree and Santee rivers and their tributaries rather than undertaking projects all over the state. One reason Wilson gave for this approach was the state's shortage of civil engineers, who would be needed to supervise the state's numerous contractors. At the beginning of this massive program, Wilson made unsuccessful attempts to recruit civil engineers from northern states:

> The employment of competent assistants to aid me in carrying on the contemplated works with expedition and economy, was deemed of the first importance; and applications were made in the different northern cities for persons who were qualified to act as Resident Engineers at the works which might be designated; but these were unsuccessful.
>
> Since the termination of the war, the attention of the different state governments have been directed to the improvement of the physical resources of the country; and the various works which are now going on in the neighbouring states, have given ample employment to those who are in any degree versed in Civil Engineering.[25]

This lack of engineers, Wilson argued, meant that the state should confined its projects to a single river system, the Congaree-Santee, which would unite the central portion of South Carolina with Charleston. Wilson explained:

> To carry on the mechanical parts of the works in digging, carpentry and masonry, either on contract or otherwise, without the aid of those who are competent to

TABLE 4.1. "Expenditures on the Rivers below the Falls, in 1819, 1820, 1821, and 1822, under the Board of Public Works"

Rivers	*1819*	*1820*	*1821*	*1822*	*Total on each river*
Wateree river under Nixon		9,373.45		292.33	9,665.78
Wateree, Congaree & Broad rivers under Barbour	24,821.34	8,333.08			33,154.42
Pee Dee river	9,803.07	7,072.09		300.00	17,175.16
Black creek		500.27			500.27
Waccamaw river			1,086.30		1,086.30
Lower Edisto	7,698.49	9,573.10	1,618.53	745.52	19,635.64
South Edisto	8,773.04	16,408.91			25,181.95
North Edisto	7,256.00	13,882.62			21,138.62
Bull's & Sullivan's Narrows		6,698.81			6,698.81
Owendaw inland passage		1,000.00	5,000.00	6,831.00	12,831.00
Santee river under Middleton		17,128.93	516.54	399.65	18,045.12
Total of each year	58,351.94	89,971.26	8,221.37	8,568.50	
Total expenditure on rivers during the years 1819, 1820, 1821 and 1822:					$165,113.07

The Wateree, Congaree, and Broad rivers under Barbour are placed under one head. The work was done in 1819, by hands hired at the expense of the State, and no account in the books point out the proportion which should be charged to each river. [The amount paid to Barbour in 1819, listed here as $24,821.34, is contradicted by *Report of the Superintendent of Public Works, to the Legislature of South Carolina for the Year 1823*, which says Barbour was paid $30,762.34.]

Source: *Report of the Superintendent of Public Works for 1828.*

TABLE 4.2. "Expenditures on Rivers below the Falls, between the 1st of January, 1823, and the 1st of October, 1828"

Rivers	*1823*	*1824*	*1825*	*1826*	*1827*	*1828*	*Total on each river*
Wateree	446.66	4,000.00		4,000.00			8,446.66
South Edisto		2,500.00					2,500.00
Pee Dee		7,000.00	1,650.00	3,445.00	1,250.00	1,945.00	15,290.00
Little Pee Dee				2,750.00	600.00	150.00	3,500.00
Waccamaw				1,718.00	500.00	500.00	2,718.00
Black River				1,144.00	600.00	200.00	1,944.00
Lynch's Creek			500.00	1,625.00	550.00		2,675.00
Saltketcher				5,500.00	1,200.00		6,700.00
North Edisto				1,400.00	1,600.00		3,000.00
Black Creek					500.00	350.00	850.00
Total of each year	446.66	13,500.00	2,150.00	21,582.00	6,800.00	3,145.00	
Whole expenditure							$47,623.66

Source: *Report of the Superintendent of Public Works for 1828.*

superintend and inspect the qualities and conversion of the materials, which are committed to their charge, would subject the state to daily impositions.

It became necessary therefore to confine to one River those operations which required the science and experience of an Engineer; and in some instance to rely upon the zeal and ability of those persons who resided near the rivers, which only required to be cleared of logs and sandbars, and who were willing to superintend the execution of those works; and in others to contract with those who were supposed competent to carry on such works advantageously.[26]

The state legislature, however, wanted immediate progress on many rivers at the same time, both above and below the falls. The shortage of civil engineers to supervise the large number of projects became a major problem for Wilson and his successors.

There were other problems as well. Information needed for planning and designing the desired waterways was almost nonexistent. Data on river flows was rarely available. River-gauging stations and stream-velocity measurements, used by today's engineers, were nonexistent. Detailed topographic studies were rarely available. It was therefore difficult for the state's engineers—as it had been for Senf on the Santee Canal twenty-five years earlier—to make informed decisions on subjects such as appropriate canal siting, desired canal depth to insure adequate depth of water in dry seasons, or the correct height of embankments and guard locks to insure adequate protection to the works during floods.

To overcome the deficiency of adequate hydrological and topological data, the engineers—as Senf did earlier—relied on the anecdotal accounts of river performance. The engineers used accounts of the drought of 1818–19 to determine expected low river levels. Similarly they used accounts of the Great Flood of 1796 to determine expected high water levels. From this information and from a visual inspection of topographic conditions, the engineers formulated informed opinions on appropriate designs for the various canals or waterways. Usually, but not always, this anecdotal approach to hydrological and topological data proved adequate. Writing of the failure of the Catawba Canal during the September 1824 flood, however, Blanding provided an example of what happened when this approach was not adequate:

In September last, this canal [the Catawba Canal] received a serious injury from the fresh in the Catawba river. This was entirely owing to the want of elevation, in the protection at its head. When the plan of this canal was settled, it was found difficult to gain correct information, as to the rise of the water at the point where it leaves the river. In the absence of this information, the great width of the river at this point, and the rapidity of the fall immediately below it, were compared with other points similarly situated, where the rise of water in the fresh of 1796 was ascertained. In making the comparison, a material fact was overlooked and led to the error of not raising the top of the guard lock sufficiently high. Since the fresh has taken place, the facts furnished by it point out the cause of this mistake. A short distance below the head of the canal, there are several islands, which prevent that rapid discharge of water in freshes which the great width of the river above, would seem to indicate.[27]

Other problems also had to be overcome, including the lack of skilled workers and cost overruns. To ease this worker shortage many were imported from the North. To attract them to South Carolina, the state had to pay higher wages, which contributed to the cost overruns.

Improving the Congaree and Santee River Waterways, 1819–1821

Connecting the rest of the state with Charleston began with work on the Congaree and Santee river waterways. To understand South Carolina's program, it is necessary to view the rivers of South Carolina as a nineteenth-century resident might. In his *Statistics of South Carolina* (1826), Robert Mills, provided such a view of the Congaree, Santee, Wateree, Catawba, Broad, and Saluda rivers, including navigation improvements undertaken up to 1826:

The Santee river enters the ocean by two mouths. There is a good steam-boat navigation on this stream, to the junction of the Congaree, and Wateree, and up both these rivers to Camden and Columbia. (The Wateree changes its name to Catawba at the Wateree Creek). This river above Camden, to the North Carolina line, is interrupted by four principal falls, around which canals have been cut, except at Rocky Mount, where the work is now going on. The first fall is at the Wateree Canal, which is five miles long, of fifty-two feet [lift], and having six locks; the second is at Rocky Mount, where there is a fall of 121 feet, requiring thirteen locks. The canal here is cut the greatest part of the distance. The third fall is at the Catawba canal, where there is a fall of fifty-six feet in three miles. The canal and seven locks are finished. The fourth fall is at Landsford, where a canal of two miles long with five locks completes the navigation. Above this the river has rapids, but the small boat navigation can be extended with care, within the Alleghany mountains. The Congaree is formed by the confluence of the Broad and Saluda rivers, where there is a fall of thirty-four feet, which is overcome by a canal three miles long, with five locks. On the Broad river the navigation for small boats extends to King's creek, with the aid of Lockhart's canal. Which overcomes a fall of fifty-one feet by seven locks in two miles. Above King's creek there are several rapids and extensive falls, locks would be requisite to make a good navigation here, and when these are once passed, the navigation to the foot of the mountains is only obstructed by a few rapids. Green river, a main branch of Broad river, extends to a point in the Blue ridge, (properly the Alleghany,) where this mountain is very low and narrow; on the opposite side of the mountain rises the French Broad, a large branch of the Tennessee. It is confidently presumed that the Atlantic and western waters may be united here by a navigable canal with great comparative ease.[28]

The principal reason for first undertaking work on the Congaree and Santee rivers was to connect Columbia, the state capital and the center of upcountry cotton production, with Charleston. Once completed, these improvements would permit canal boats to proceed down the Congaree and Santee rivers from Granby and pass through the Santee Canal to Charleston. Steamboats and larger riverboats that were too big to fit through the Santee Canal could proceed to the mouth of the Santee River and thence by the ocean to Charleston.[29] These improvements would decrease the cost of shipping cotton from central South Carolina and increase the amount of cotton shipped. They would also decrease the cost of shipping salt and manufactured goods from Charleston to the central part of the state.

Once the waterway from Granby to Charleston had been improved, opening the Saluda River to navigation would facilitate the transportation of cotton and other agricultural produce to Charleston from the western part of the state. Opening the Broad River to navigation would facilitate transport of cotton to Charleston from the north-central part of the state. Cotton shipped down the Saluda and Broad rivers would be carried past the rapids at Columbia by a new canal, the Columbia Canal (see color plate 8). It was thought that the Saluda River navigation, as well as improvements to the Edisto River, would divert trade from the Savannah River and Charleston's chief competitor as a port city, Savannah.[30]

We have been politely favored with the following extracts from a letter received by a gentleman of this City, from Major JOHN WILSON, State Engineer, dated

"*Col[illegible]ia, September 22.*

"I think it is an erroneous system to expend monies on the upper parts of our streams, while the lower retain all their obstructions. With this view of the subject I have invariably directed my attention to the removal of those obstacles which embarrassed the Navigation from the Fork of the Watteree and Congaree to this place. The completion of the navigation to Columbia, will enable me to obtain from the Sea Board, with less difficulty and expense, all those heavy articles which are absolutely necessary for our works on the Broad and Saluda.

Southern Patriot, September 26, 1818

The principal obstructions to navigation on the Congaree and Santee were sandbars and logs. Clearing these rivers began in September 1818, when local citizens donated their workmen, presumably slaves, to the state engineer for clearing the river.[31]

Beginning in 1819, voluntary efforts were superseded by those of contractors. Mordecai Barbour of Virginia was the first contractor to undertake work on the Congaree and Santee. Progress was slow. Wilson reported: "In the month of March [1819] he [Barbour] brought with him to Columbia Thirty-three men; and in the first week in May only one set of flats was prepared for the Congaree river. About the latter part of July two other sets were floated down the Congaree, and began in August, at the mouth of Wateree river; [but] by the tardiness of these preparations the best part of the season was lost. The first set of flats cleared the channel of the Congaree from Granby to Haig's Neck, which is by land about twenty miles from Columbia. In August they dropt down to the mouth of the river, and since then have been working upwards; on the Wateree they have ascended to Bently river, and on the Congaree to Butler's Neck."[32]

In addition to his work on the Congaree and Santee, Barbour also did some river clearing on the Broad and Wateree rivers. For the work on all four rivers, Barbour was paid $30,762.34.[33]

The contractor also built structures, called jetties, to improve navigation. These jetties were intended to swell the waters of the Congaree so as to permit the passage of boats, particularly at times of low water. Two jetties were built out into the Congaree: one 150 yards long located a half mile below Granby and the second 250 yards long ten miles below Granby.[34]

Barbour's contract was not renewed in 1820, and the work on the Congaree and Santee was continued that year by a new contractor, Colonel W. J. Myddleton: "In April [1820] Col. Myddleton began the work on the Congaree where it left off last year, and has descended some distance down the Santee. The only formidable difficulties now existing in the navigation from Granby to Charleston, are the shoals in the upper part of Santee and the rapids at Murray's ferry.[35] A more favourable state of the river would have enabled Col. Myddleton to remove these before winter. When the river is low a few months will complete it. The narrows at Bull's Bay are finished, and those at Sullivan's Island nearly completed."[36]

While Barbour and his men primarily removed logs and sandbars from the Congaree and Santee, in 1820 Colonel Myddleton and his men began cutting channels around the rock obstructions remaining in the Santee River above Gourdin's (Murray's) ferry.[37] This work was done by slaves. Blanding's instructions to Myddleton, which still exist, specified that Myddleton should use fifteen slaves per flatboat to remove river obstructions.[38] Blanding also specified the rations that these slaves should receive: "Slaves in the Flats shall be allowed three quarters of a pound of bacon, or a pound & a quarter of Beef per day, and at the rate of one peck of meal & two quarts of peas per Week, and one gill of Whiskey or molasses per day."[39] Contrary to some assertions that slaves were underfed on the South Carolina projects, comparison of Blanding's instructions to Myddleton to similar documents relating to northern canals indicate that the slaves were fed in the same quantities, and were provided the same amount of whiskey, that the northern canals gave their white workers.

Initially Myddleton's progress was slow. In 1820 he and his men were able to work on the Congaree-Santee navigation for only two and a half months owing to heavy rains and the early onset of the sickly season.[40] This work was enough to open the Congaree River to navigation except for the sandbars eleven miles below Granby and those at Barbour's cut twenty-one miles below Granby. These sandbars could be passed by boats during most of the year except the very dry season.[41]

The Congaree River begins at Columbia, where it is formed by the joining of the Saluda and Broad rivers. River navigation from the Broad River to the Congaree River at Granby was not possible because of the falls at Columbia. The Columbia Canal was later constructed to bypass these falls.[42]

Work continued on the Congaree and Santee for several years, primarily to make low-water navigation safe for steamboats.[43] By 1819 enough improvements had been made along the Congaree and Santee to allow steamboat operations to

Granby: "With the improvements already made, a good steam boat navigation is opened between Charleston and Columbia, which has reduced the price of freight one half. The further works to be done there will render it quite safe and expeditious."[44]

Opening the Saluda River Navigation: Saluda Canal, Drehr's Canal, and Lorick's Canal, 1819–1826

While work was being conducted to open the Congaree and Santee, work was also begun on the Saluda River. Though State Engineer John Wilson wanted to open Congaree and Santee rivers below the falls first, then turn his attention to building the Columbia Canal around the falls of the Congaree and later undertake navigation improvements on the Saluda and Broad rivers, his plan was not to be. State legislators from the upstate did not want to wait for improvements in their part of the state. Political pressure dictated that Wilson start working immediately on the Saluda and Broad rivers above the falls at Columbia.[45] The legislature specified that work on the Saluda Canal, immediately west of Columbia, begin as soon as possible. As in other waterway-improvement projects, the Saluda River navigation was to be a combination of bypass canals around the most substantial obstacles and river sluices and obstacle clearing in the rest of the river (see color plate 9).

The Saluda Canal was the first of three bypass canals constructed as part of the Saluda River navigation. It entered the Broad River immediately upstream from the confluence of the two rivers, providing passage around the rocky shoal at the mouth of the Saluda River. This shoal extended three hundred yards upstream from the mouth of the Saluda and had a fall of nine feet. The canal also bypassed Beard's shoals a mile and three quarters further up the river. These shoals extended two hundred yards along the river and had a fall of fourteen feet. A mile further up the Saluda River, the canal also bypassed the shoals at Senn's. These shoals extended for a mile and a half along the river and had a fall of thirteen and a half feet.[46] The upstream end of the Saluda Canal began just above Senn's shoals, and the canal extended a little over two and a half miles.

On January 25, 1819, Wilson entered into a contract with Captain Thomas Strode to build the Saluda Canal. In March 1819, while Mordecai Barbour began river-clearing operations on the Congaree and Santee rivers below Granby, Strode began work on the Saluda Canal. Wilson informed the governor that a part of his workforce was then engaged in opening the canal around the rapids at Senn's and Beard's and that, once the work was completed, the Saluda would be navigable for twenty miles[47] (see figs. 4.1 and 4.2).

Captain Strode, however, encountered difficulties in recruiting the required workforce. In August of 1819 his contract was assigned to his creditors, and Barbour took charge of the work.[48]

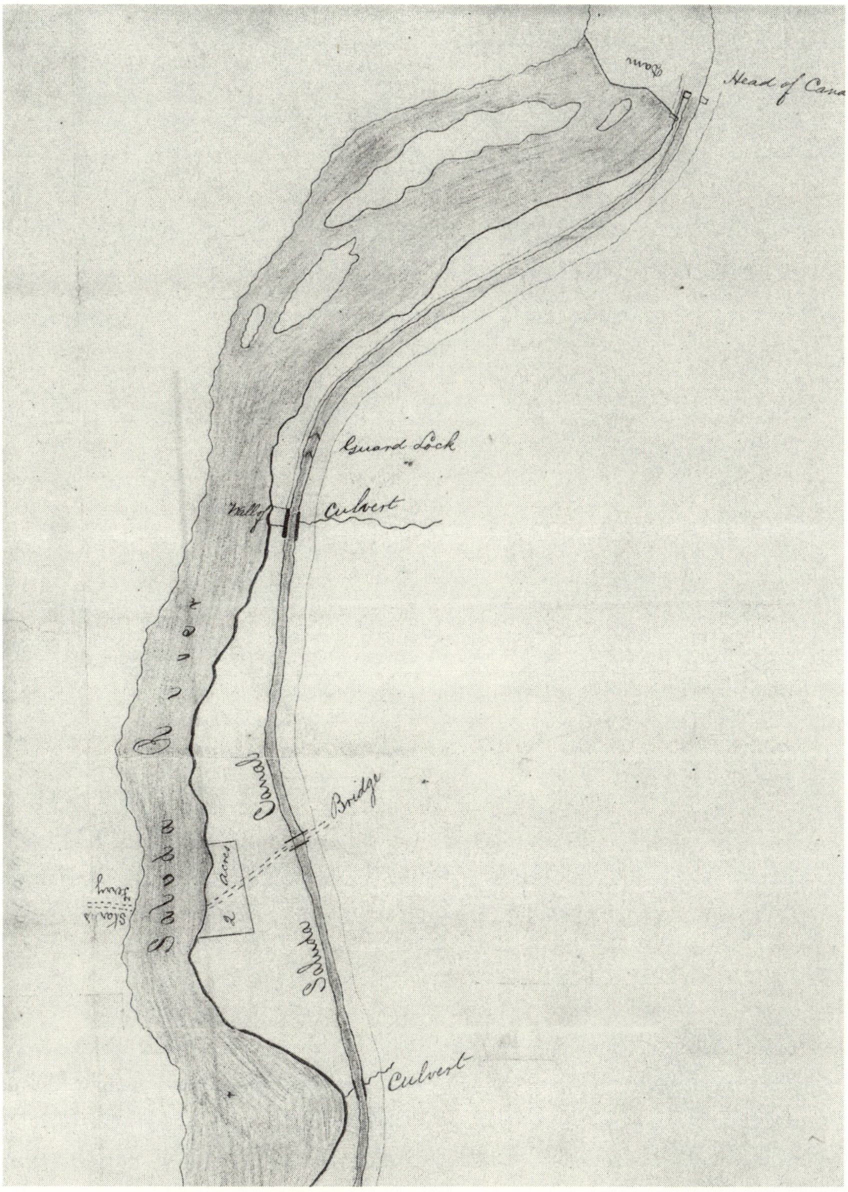

Fig. 4.1. "Plat of Land of A. B. Starke taken by the Board of Public works for the Saluda Canal from Mr. Starke's Upper tract," ca. 1839. Upper Saluda Canal, 1839. South Carolina Division of Archives, Department of Public Works; Superintendent, Road, River and Canal Projects File, 1820–1845 [S208009]. West is at the top. This plat is one of a very few extant drawings depicting the canals constructed by the Board and Superintendent of Public Works. Senn's shoals (not labeled) would be immediately downstream of the dam across the Saluda River (upper right). The guard lock (center) is not located at the head of the canal, where one might expect to find it. Starke's Ferry and the bridge constructed by the Board of Public Works are in the center.

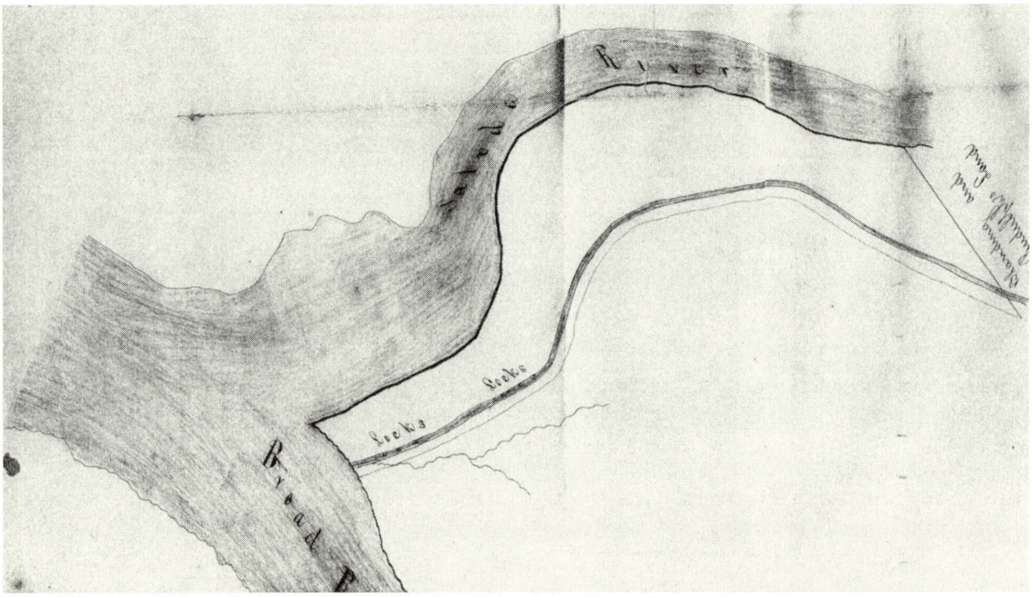

Fig. 4.2. "Plat of Land of A. B. Starke taken by the Board of Public Works for the Saluda Canal, from Mr. Starke's lower tract," ca. 1839. South Carolina Division of Archives, Department of Public Works; Superintendent, Road, River and Canal Projects File, 1820–1845 [S208009]. South is at the top. The lower Saluda Canal is shown paralleling the Saluda River (top) and entering the Broad River (left). The Broad River dam and the entrance to the Columbia Canal are not shown.

By the beginning of 1820, much of the work on the Saluda Canal had been completed. In that year the Board of Public Works provided the state legislature with a description of the Saluda Canal and the progress to date:

> When the Board commenced operations in February [1820], the Saluda canal had been nearly excavated for two miles and a half.[49] At the head of this canal a large extent of very hard rock remained. This has been blasted out, and a guard lock erected to protect the canal from, the very high freshes of this river. Two locks of eight feet lift have been built of hewn granite, and all materials collected for the other two. It will require not more [than] two months to finish these locks: and but for the suspension of the work by the early sickness which drove the workmen from the rivers about the first of July, they would have now been completed. It is expected this canal will be finished by January [1821] over-coming a fall of thirty-five feet which now closes that river, and precludes the possibility of a boat's passing. This navigation will be extended at least ninety miles above Columbia.[50]

In addition to four masonry lift locks and one stop lock on the Saluda Canal, the contractor and his men also constructed

---

A letter from JOHN WILSON, Esq. civil and military Engineer to the State, dated Columbia, 16th inst. to His Excellency the Governor, states that part of the workmen employed by the State, are engaged on the Saluda river in opening Canals around the great falls at Senn's and Beard's; and that the Saluda will be navigable for 20 miles when the work is completed.

Southern Patriot, March 25, 1819

a timber crib dam across the Saluda River at the upper end of the canal. It was four feet high and seven hundred feet long.[51] Also constructed were four culverts and two waste weirs. The contractor built a masonry bridge on the road to the Saluda (Stark's) Ferry. This road, which passed over the canal bridge and across the Saluda Ferry, proceeded southeasterly to Granby below Columbia. Above the Saluda River, the road proceeded northwesterly to the upcountry. It was the only crossing of the Saluda River for a considerable distance. Unfortunately the bridge was not well built. By 1826 it had become dangerous to users and had to be taken down and rebuilt.[52]

The Saluda Canal was completed and opened to navigation in 1821. It had taken three years to complete.[53] The 2.59-mile canal cost $150,466.73, or about $58,000 per mile—substantially more than the cost per mile of the Santee Canal. Once the Saluda Canal had been opened, navigation was possible from Columbia, across the Broad River, through the Saluda Canal and up to Drehr's Mill and Ford—a distance of about twenty river miles.[54]

With the completion of the Saluda Canal, the Board of Public Works turned its attention to extending the Saluda River navigation westward. As early as 1820, the board realized that the Saluda Canal would not be useful until a bypass canal was built at Drehr's Mills: "As this part of the navigation will not be useful, until the obstructions above [the Saluda Canal] are removed, it was thought expedient to extend the works immediately to the shoals, at Drehr's mill, where there is a fall of twenty-one feet in eleven hundred yards."[55]

To open the rest of the river to navigation would require a bypass canal fourteen land miles above the mouth of the Saluda River at Drehr's Mill, where the river dropped twenty-two feet in eleven hundred yards,[56] and a smaller, less expensive bypass canal at Lorick's Mill fifty land miles above the mouth of the Saluda River.

At Drehr's Mill the board decided to build a canal eleven hundred yards long (0.63 miles) with one guard and three lift locks to bypass the mill and overcome a fall of twenty-one feet. Work was begun in 1820 by a contractor named Mr. Kibbe.[57] Besides the lift locks and canal prism, construction projects included a stone lockkeeper's house, an aqueduct, several culverts, and other structures.[58]

At Lorick's Mill the board eventually decided to construct a small bypass canal instead of replacing the existing mill. The canal was to be one mile long with one lock. The lock was to be constructed of wood instead of masonry and was to overcome a fall of eight feet.[59] Although this lock no longer exists, Blanding's description of this wood lock has survived.[60]

Construction on Drehr's Canal began in April 1820. The board reported to the state legislature on its progress: "The excavation in part, through very hard rock, has been finished. The stone culverts are completed, one of which is so large as to pass a creek under the canal. The stone for the locks are partly cut out. Messrs. Dyer, Gass and Robinson will have a competent force on this work by the middle of October."[61]

The board was very pleased with the progress of contractors Dyer, Gass, and Robertson. In their 1821 report to the legislature, they wrote: "The locks on this canal have been completed, and altho' they have not yet been tested by passing boats through them, yet the Board have great confidence that the work has been faithfully executed. When this work was undertaken, the contractors, Messrs Dyer, Robertson and Gass, were not expected to finish them this year."[62]

The board was not as pleased with the work done by their other contractor working on the Drehr's Canal, Mr. Kibbe. They reported to the state legislature: "The execution of the guard lock, aqueduct, culverts, and canaling [excavation of the canal prism] were performed, under another contract with Mr. Kibbe. A part of this work was so badly done, that it was not received by the Board, but rejected. Their decision of other parts of the work, has been suspended, until the canal can be put in operation and the work tested."[63]

It was this project, as well as the initial construction of the State Road, that brought great criticism upon the Board of Public Works. Critics asserted that the board had not exercised sufficient supervision over the contractors—one of the outcomes of not having enough civil engineers, as Wilson had warned. But the board also had its supporters. One editorial lauded its work: "That the works on the other Canals, until suspended by the sickly season, have been actively carried on & well executed, except at Drehr's Mills, on the Saluda, where the defects were pointed out by the commissioners and payment refused."[64]

In his 1823 report to the legislature, Blanding announced that Drehr's Canal had been completed in 1821 at a cost of $79,139.65.[65] But there was another problem: Drehr's Canal served the goal of opening up the entire length of the Saluda River navigation only if the other work along the river, particularly Lorick's Canal at Lorick's Mill, was also finished. In 1824 Blanding had to report to the legislature that Lorick's Canal had not yet been completed: "As the works above [Drehr's Canal] have not been finished, this canal has not been used and no tolls have been established on it.... As the Saluda will be navigable to the Abbeville line, as soon as the lock at Lorick's mills is finished, this canal will come into use next spring, if not earlier."[66]

At Lorick's Mill, Blanding had entered into an agreement with the mill owner, Jacob Lorick, to build a canal around his mill dam. Lorick agreed to advance three fifths of the cost, which was later to be refunded to him by the state:

> But still the obstructions to the navigation of this river, presented by Mr. Jacob Lorick's mill-dam remained. To have opened this dam to an extent which would have rendered the sluice through it passable by loaded boats, would have been attended with considerable expense, and would have rendered Mr. Lorick's mills, which have been erected at great expense, totally useless; thereby occasioning a great and serious loss to Mr. Lorick, and a great public inconvenience to the country around, especially

on the Edgefield side of the river. The ground was found very favorable for locking around this dam on the north side of the river, and when effected the navigation would be vastly better than any sluicing could render it. I had a consultation with Mr. Lorick, and we agreed that a canal and a wooden lock should be constructed there, on a plan of durability, in which I had confidence. I obtained a contract for this work, with Mr. John S. Walsh, for $2,500, agreeing on my part to find the irons for hanging the gates. On the part of the state, I agreed to pay one thousand dollars, and Mr. Lorick, agreed to advance fifteen hundred dollars, relying that when all the circumstances should be submitted to the legislature, the state would refund the amount to him.[67]

The contractor, John Walsh, vigorously pursued work on the canal and lock. Although his work was seriously damaged in the September 1824 flood, delaying its completion,[68] by March 1825 the lock at Lorick's Mill was completed.[69] Yet Lorick's Canal continued to be a problem. In 1826 Blanding reported: "At Lorick's dam, the executors of the estate of his estate have made a new arrangement. To avoid flowing [flooding] the lands above, and to permit the passage of fish up the river, they have lowered the dam two feet, and thereby reduced it to the elevation it had when it was first built, about forty years ago. This rendered it necessary that the short canal here, which communicates with the river above the dam should be deepened two feet, and that the breast of the lock should be lowered as much."[70]

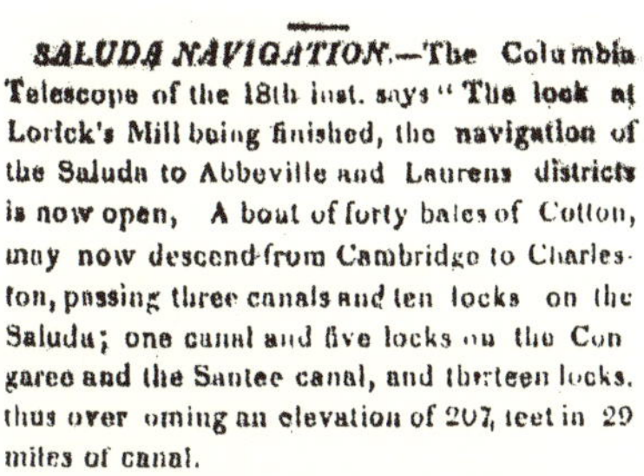
SALUDA NAVIGATION.—The Columbia Telescope of the 18th inst. says "The lock at Lorick's Mill being finished, the navigation of the Saluda to Abbeville and Laurens districts is now open, A boat of forty bales of Cotton, may now descend from Cambridge to Charleston, passing three canals and ten locks on the Saluda; one canal and five locks on the Congaree and the Santee canal, and thirteen locks. thus overcoming an elevation of 207, feet in 29 miles of canal.

Southern Patriot, March 22, 1825

Besides three bypass canals, Blanding also had to construct sluices on the Saluda to make the river navigable. In 1823 Blanding hired Colonel Moman and his fifty-four-foot-long, eight-foot-wide Broad River boat to ascend the Saluda River and investigate the remaining obstacles. Blanding left Columbia with Colonel Moman, board member Robert G. Mills, and several other gentlemen to investigate the state of navigation along the river. Blanding then submitted a detailed report on the Saluda River navigation to the state legislature:

The following is the state of the river as examined at common winter water. In summer, it will be usually found lower, but then boats will rarely run, even when the water is up.

From Columbia to Hyler's shoals, a distance of 18 miles by water, the canals [the Saluda Canal] have opened a good navigation for boats carrying 100 bales of cotton.

Hyler's shoals are about half a mile long, very gradual, with deep water, and easily passed.

Smoke's shoals begin immediately above these, and extend about one and a half miles; the fall is about 17 feet on a regular inclined plain. The bed of the river is very wide, and on the several ledges which cross it, there is generally a want of water. In most of them, there are sluices having sufficient depth, but in consequence of their not being opposite to each other, the boat channel is now difficult. To render this shoal easy for the passage of boats carrying 35 bales of cotton, it is necessary that a sluice twelve feet wide should be opened through six ledges, and that two or three small wing dams should be constructed to swell the water in the boat channel. This shoal is the most difficult now remaining to be opened.

Hamiter's shoals are 4 miles above Smoke's. They are about a half mile long, and fall about 5 feet. The water is deep, and only interrupted by two ledges, neither of which are difficult, but require some work.

To Snelgrove's island, the navigation is good. Here the river is wide and shallow, and falls gradually about 6 feet in a half mile. By extending a wing dam from the upper point of the island towards the south shore, and clearing out the channel under the left bank of the river, this navigation will be safe and easy.

Two miles above are the shoals of G. Drehr's mill, which fall about 5 feet in 600 yards. The water is deep in the south channel, and a very little work is required to render it good.

At Simmon's ferry, there is a gradual fall of 7 feet in ¾ of a mile; the water is shallow, and a few wing dams will be required to keep up the proper depth. There must be several rocks blasted to give the channel a proper direction.

The shoals at Dominick's mills are 8 feet in one mile, with deep water. The dam of this mill is broken, and there is a good sluice through it on the north side of the island.

Above Water's ferry, in Newberry district, there is a small shallow shoal requiring the sluice to be deepened, which may be easily done.

Between Waters and Bates, there is a gradual fall of seven feet in 600 yards. There are five ledges passing the bed of the river, through which a sluice should be opened.

At Wright's mill, the fall is four feet in 503 yards. On the south shore, the water is deep, and interrupted only by the remains of an old dam, which may be easily removed.

Between Wright's and the mouth of Little Saluda, there is a gradual shoal of four feet in one mile. In one part of which, some work is required to gain a sufficient depth of water.

Between the Little Saluda and Lorick's mill, there is a small shoal of 2 feet fall in 200 yards, where the channel requires to be cleared for a short distance.

Lorick's mill dam extends entirely across the river, and is six feet high. The current in this part of the river is so gentle that the dam makes back water above it for 7 miles. The removal of this obstruction, which is entirely artificial, is all that is required. To canal and lock round this dam would cost four times the value of the mills. The extent of boat navigation above this point, capable of being rendered perfectly good, with an expenditure of from $2000 to $3000, is at least seventy miles by the river, so that this obstruction to the navigation ought not to be suffered to remain.

From Lorick's to Mason's mills, two miles above the lower line of Abbeville district, the distance is 42 miles by water, which we ascended in less than two days, and returned in one,—In this whole distance, there is but one shoal requiring any work. This is at Robertson's mill, where the fall is about four feet in 300 yards. The water is deep, and a sluice on the north shore would cost but little, and make a safe navigation. The river this distance is narrow, deep and lined with rich low grounds, presenting in every respect, the appearance of our rivers below the falls.

At Mason's mill dam, we stopped. Archibald Mason, Esq. the owner of the property, expressed his perfect willingness to take down the dam at his own expense, as soon as boats shall begin to run on this part of the river.

My report on the river, between this dam and Ware's mills, a distance of 26 miles by water, is made from information and from examinations previously made by me. From Mason's to Young's mill, the distance is 13 miles, without any material obstruction. Here there is a ledge across the river, with deep water above and below it, so that the shoal is capable of being made perfectly good.

Adams's falls are the only remaining obstruction of any difficulty between Young's and Ware's. These falls are caused by two ledges, through one of which there is now a good sluice, and very little labor is required to make a good one through the second. Two other small shoals, of no difficulty, intervene between this and Ware's, where the navigation of the Saluda terminates.[71]

The Saluda River navigation was impressive. It extended 118 river miles or 95 land miles from Ware Shoals, immediately at the Greenwood County–Abbeville County–Laurens County line, all the way to Columbia.[72]

Problems still remained in the Saluda River navigation. As late as 1826 Blanding was reporting, "the sluices were therefore not sufficiently open on the principal Shoals, and at several minor ones, it was different [difficult] to pass in low water."[73] Despite this, Blanding was able to report to the legislature the opening of the Saluda River navigation: "Yet, with these impediments in the way, boating commenced on that River last winter, and furnished the evidence that the stream possesses the capability of great usefulness."[74]

The improvements to the Saluda River had been expensive. Blanding was aware of this high cost but believed it was justified: "Expensive as have been the Saluda and Drehr's canals, I think this view [the resulting savings of twenty-four thousand per year on eighteen thousand bales of cotton shipped from the Saluda River valley] ought to satisfy every one, that by unlocking the Saluda river, they are not to be regarded as a waste of public treasure."[75]

Building the Broad River Dam, 1819–1823

The Saluda River navigation would not be useful until descending boats could cross the Broad River and continue through the Columbia Canal around the falls at Columbia to Granby, the beginning of the Congaree-Santee passage to Charleston (see color plate 10). A dam had to be built across the Broad River above Columbia to permit boats to cross from the Saluda Canal to the Columbia Canal. This dam would have four functions: (1) to increase the height of the water in the lower end of the Saluda Canal and thus avoid building expensive locks to bring the boats down from the level of the Saluda Canal to the Broad River; (2) to increase the height of the water in the Broad River to facilitate passage of boats from the Saluda Canal; (3) to provide a slack-water navigation for the lower end of the Broad River; and (4) to throw water into the Columbia Canal, yet to be built.[76] The estimated cost of this dam was twenty thousand dollars.

The Broad River Dam was substantial. It was a timber crib dam approximately fifteen hundred feet long, and it rose six feet above the twelve-foot-deep waters of the Broad River. Typical of crib dams built in the nineteenth century, the dam was anchored into the riverbed by timber bolts installed during times of low water. On top of these timber anchors was built a timber box or "crib," into which workers dumped rock and soil. The rock and soil were sealed against leakage by tamped clay known as "puddling."[77] Puddling was a nineteenth-century construction technique of lining the bottom and sides of the canal with an impervious layer of clay to forestall leakage. Liquid clay was tamped into place and formed the impervious layer.

In 1823 Blanding provided a description of the dam for the state legislature:

> This work [on the Broad River Dam] which has been so often obstructed, and so long delayed by the frequent freshes of the last two years, was finished during the last summer [1823]. . . . This dam is built six feet above the surface of the water, which in some parts of it, is 12 feet deep. It is 1500 feet long, being divided into four parts by three islands rendering eight high stone abutments necessary. These islands are of light alluvial soil and there was a necessity of puddling the backs of the abutments with clay boated some distance. Before this could be done, in several places freshes passed behind the abutments, and formed extensive openings. These have all been filled with clay the last summer, raised above freshes, and are now safe.[78]

The Broad River Dam was unusual among nineteenth-century dams in one respect—a fifteen-foot-wide fish sluice was built into the dam to allow fish to migrate upriver. The Board of Public Works included a provision for a similar sluice in the Saluda Dam contract.[79]

The contractor for the Broad River Dam, a Mr. McKenzie, began work on the dam in 1820 "with a respectable [workforce] and [a] considerable impression has already been made on this work."[80] The work was disrupted by the flood of September 1821, the largest such flood on the Broad River since the Great Flood of 1796. Smaller floods throughout the winter of 1821–22 also hindered the work.

McKenzie and his men finally completed the dam in 1823. Unfortunately McKenzie died of drowning on the dam the day after it had been reported completed: "The indefatigable industry and perseverance of Mr. McKenzie, the contractor, would seem to have entitled him to a better fate. He was drowned on the spot which had been the scene of his labors, the day after they were brought to a close."[81]

The demise of the "indefatigable Mr. McKenzie" on the dam one day after its completion made a compelling story. In fact the dam was uncompleted at that time and leaked badly. Blanding reported the situation to the state legislature: "During the last summer, it was rendered evident that the Broad River dam, although strongly and substantially built, had not been entirely finished when the contractor was drowned. For in the very low water of last August, the whole river passed through the leaks of the dam (which is 1500 feet long) and left the top of it bare, and thus subject to decay."[82]

Blanding worked with the administrator of McKenzie's estate, and remedies were made to the dam. The September 1824 flood showed additional defects in the masonry in the dam's abutments. These too were corrected,[83] but problems continued with the Broad River Dam. In 1826 Blanding reported that leakage through the dam in that unusually dry year deprived the Columbia Canal of sufficient water for navigation:

> To the last Legislature, I reported the state of the dams connected with this work [Columbia Canal and Bull Sluice Canal]. Since that period, the season has been unusually dry, and the defects in these dams have become manifest. They were erected in years of unusual rains, and were then considered sufficiently tight to retain the water necessary to feed the canal connected with them: But this year of remarkable low water, the contrary has become apparent.—When Broad river becomes very low, as all our rivers have been this year, the leaks in these dams are so great, that a sufficiency of water is not supplied for the canal navigation. This produces effects injurious to the interest of a large section of country dependent on it for transportation, and of the town of Columbia; and at the same time diminishes the income from tolls. This was so manifest, that I undertook, during the summer, to have stone and gravel boated to the dam, and thrown on the upperside of it; but was not able to complete the work without incurring a larger expense than I was disposed to incur on my personal responsibility.[84]

Although the leakage of the Broad River Dam limited water flow to the Columbia Canal, it did not stop boat traffic totally. Blanding reported that 18,968 bales of cotton passed through the canal in the first three quarters of 1826. Had the Broad River dam not leaked or had 1826 not been such a dry year, he estimated that more than 30,000 bales would have transited the canal.[85]

Improvement of the Broad River Navigation: Bull Sluice Canal and Lockhart's Canal, 1820–1822

The state legislature had directed that the Broad River also be promptly improved for navigation. Unlike the Saluda River, the Broad River above Columbia did not have extensive obstacles and was already being used for navigation, usually during the high waters of spring. A typical large Broad River boat operating on the river at this time was fifty-four feet long and eight feet wide with a draft of eighteen inches. Crew for such a boat was usually four oarsmen and a helmsman. The boat typically had a capacity of sixty bales of cotton.[86]

The principal obstacle on the river was Bull Sluice at Guignard's Island, one and a half miles above the confluence of the Broad and Saluda rivers. Bull Sluice was a natural channel between the island and the east bank of the Broad River, but because of the gradient, travel through the channel could be perilous. The Board of Public Works directed the construction of a still-water canal parallel to and adjacent to Bull Sluice. This new canal had an eight-foot lift lock and an eight foot-stop lock (see fig. 4.3).

McKenzie, the same contractor who was building the dam across the Broad River, began work on Bull Sluice Canal in 1820. In its 1820 report, the Board of

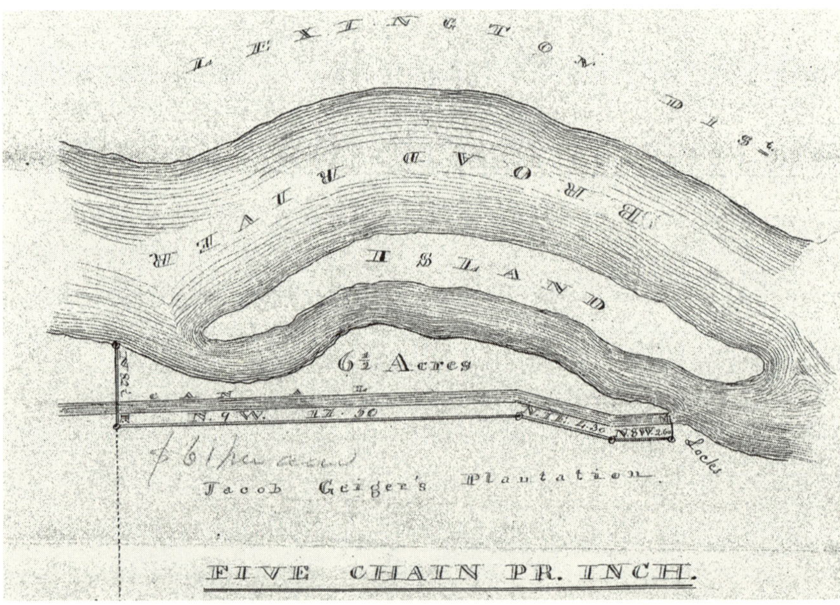

Fig. 4.3. Plat showing Bull Sluice Canal and the Broad River, May 1839. South Carolina Division of Archives, Department of Public Works; Superintendent, Road, River and Canal Projects File, 1820–1845 [S208009]. West is at the top. The island is Guignard's Island. Bull Sluice is located between the island and the canal. The locks on this canal (right) are at the north end.

Public Works was able to tell the legislature that the foundation and coffer dams for the bypass canal were under construction.[87]

By 1821 McKenzie and his men were making good progress on finishing the stop lock, lift lock, and wing wall at the head of Bull Sluice Canal. Then disaster struck in the form of the massive September 1821 flood. The board reported the damage to the legislature:

> The Lock at these falls [Bull Sluice] was finished early in the summer [1821], and the embankment connecting it with the island, was raised within a few feet of the elevation requisite to place it above all freshes. This was the state of the work, when the sudden and unexampled rains in September raised the Broad and Saluda higher than they have been known since 1796. The rise of water was unusually rapid, having from a low river, gained its greatest elevation in the course of one night. It passed a few inches over the top of the embankment, and the whole was swept away before morning. The Lock and embankment next to the mainland remained uninjured. As soon as the water subsided, the contractor began reconstructing this embankment, but it is not yet finished. The season has been particularly unfavorable to this work from its exposed situation. It was several times entirely covered with water before the lock was completed, but the faithful execution of the masonry resisted the violence of these torrents.[88]

By 1822 McKenzie and his men had repaired the damage, and Bull Sluice Canal was opened to navigation.[89]

There were other obstructions to navigation in the Broad River above Bull Sluice, but none was as severe. In 1818 Major Wilson reported to the state legislature: "The obstructions to the navigation of Broad river, commence from its mouth at Young's Mill-dam, by a rocky shoal two hundred yards in length, and with a perpendicular rise of nearly three feet eight inches. The next obstruction is at Curries Island, where is a rapid of two hundred yards, with a rise of one foot. The river has been carefully examined twenty-five miles above Starke's ferry. In this distance there are fifteen obstructions to its navigation. From the Bull Sluice, the first immediately above the ferry, to a point nearly opposite Crane Creek, the perpendicular rise of the river is about 8 feet."[90]

It was Wilson's plan (and later the plan of the Board of Public Works) to bypass these obstructions with sluices:

> Ascending from Crane Creek [immediately above Bull Sluice, entering Broad River on the east bank], the current is gradual for two and a half miles, to Rive's Mill, where the rise from the Island opposite the Mill-dam to the upper part of the shoal, is seven and a half feet. From hence, a regular current for two miles, to the third obstruction. From No. 3, to the shoal above Williamson's, the perpendicular rise is five and a half feet. At the lowest end of the Island, above Williamson's, a ledge of rock

extends across the river; the rise to the upper part of the Island is five feet. From hence for a short distance, the river is navigable. Obstructions No. 5 and 6, have together a rise of three feet nine inches. Obstruction No. 7, occurs a little above Cedar Creek, with a perpendicular rise of nearly three and a half feet. From this point to the lower part of Rives's Island, about three quarters of a mile, the current is gentle, from here to the upper end of the Island, the rise is nearly four and a half feet. Three quarters of a mile above Rives's Island, the river is divided by a cluster of Islands; from the lowest of them to a small Island above Patrick's Creek, there is an extensive rapid, with a perpendicular rise of twelve feet. From hence to a point above Little River, there is a rise of twelve feet more; and from this point to the eleventh obstruction, the rise is three and a half feet. The twelfth obstruction commences at the fish-traps below a large Island, and extend to Freshly's Mill-dam, having a rise of eleven and a half feet. The current is regular from this to about one mile above Chappell's, where there is a rapid with a perpendicular rise of two and a half feet.

At Hall's fall, about a mile higher up, the rise is two and a half feet; from Halls to Cooks, the obstructions are inconsiderable; and between Cooks and Longs, there is a rise of fourteen inches. From this point to the mountains, the river has not been minutely examined, but the nature of the obstructions is the same with those already described.[91]

Work began in June 1819, primarily by cutting sluices in the bed of the river.[92] Wilson, however, was not pleased with the contractor's work and suspended contract operations on August 31, 1819.[93]

In the following year, the new Board of Public Works continued "opening channels in the bed of the river, wherever this method of improving its navigation was practical."[94] Unlike the Saluda River navigation, construction operations on the Broad River were relatively easy. This encouraged the Board of Public Works to begin work on Lockhart's Canal, at the very northern portion of the Broad River passage. "The extensive water communication afforded by this stream [the Broad River], with some of the finest provinces of North Carolina, and with an important section of our own state, together with the advantages to be derived from the limestone and marble quarries situated on its banks, as soon as those heavy materials can be transported by water, induced the Board to extend the works on that river."[95]

Lockhart's Canal, also called the Broad River Canal, was located at Lockhart's shoals, seventy-five miles above Columbia, northwest of Chester (see color plate 11). This bypass canal was 1 mile and 2,376 feet (1.45 miles) long and consisted of one guard lock and six lift locks, all masonry, to overcome the 47-foot fall of the river. It was intended to replace the portage the state legislature authorized in 1813.[96] The locks were 10 feet wide, and there were 70 feet between gates. Begun in March 1820 by contract with Colonel Wilson Nesbitt, the canal prism excavated was 12 feet wide at the bottom and contain 2.5 feet of water.[97]

Blanding described to the state legislature the situation at Lockhart's shoals: "Lockhart's shoals, where this canal [Lockhart's Canal or the Broad River Canal] is constructing, extend one hundred and ten chains [7,260 feet, or 1 mile and 1,980 feet (1.38 miles)].[98] They are divided into two parts, by a sheet of deep water. The upper part has a fall of thirty-four feet, and the lower part of thirteen feet. For about fifteen chains [990 feet], at the lower end of these shoals, high hills, filled with the very hardest rock, rise immediately from the water's edge. These hills present the only very formidable difficulty in effecting this navigation."[99]

The Board of Public Works looked at four different plans for a canal around Lockhart's shoals. Writing for the board, Blanding described the plan chosen: "The first plan proposed to continue the level, about sixty chains [3,960 feet or 0.75 miles], to a ravine, then to descend by two locks to a new level, which should be continued to the upper point of the hills, and there descend into the river, by three locks, fifteen chains [990 feet] above the foot of the falls. In these fifteen chains, there are five feet fall, which was proposed to be overcome by a wing dam, and one lock at the foot of the falls."[100] A lockkeeper's house was to be built on a hill overlooking all of the locks.[101]

The contractor, Colonel Nesbitt, began work on the canal in March 1820. By the end the year, the board was reporting to the state legislature: "The digging has progressed about three-fourths of a mile. The foundation of the guard lock has been prepared, and the materials for it collected. Stone for the other locks of the very best quality has been discovered, and a part of the aisler [ashlar] for locks Nos. 1 and 2, has been cut."[102]

By the end of 1821 the board could report to the legislature: "The Canal at Lockharts shoals has been mostly excavated, but not finished off. There have been three locks built and another commenced. The dressed stone for the remaining work, is much of it prepared. There has been one bridge erected over this Canal, which is the only one required on it."[103]

The Board of Public Works had problems with the contractor's workmanship on two of these locks. The work was found inferior and had to be redone.[104] By 1824 the canal was completed, opening the Broad River navigation to the area above the present-day town of Lockhart in Union County.[105] The total cost was $48,409.40, paid to contractor Nesbitt.[106]

Building the Columbia Canal, 1820–1824

In the second decade of the nineteenth century, the State of South Carolina made substantial progress in opening the central portion of the state to water transport. By 1819 improvements on the Congaree and Santee rivers had permitted steamboats to reach Granby from Charleston. By 1821 the recently constructed Saluda Canal had made the Saluda River, extending west from Columbia, navigable for 20 miles. Additional improvements were undertaken to extend navigation on the

Saluda River 110 miles from Columbia. The Broad River, extending north from Columbia, was open to improved navigation because of the canal at Bull Sluice, opened in 1821 and repaired of flood damage by 1822. In 1823 Superintendent of Public Works Blanding continued this progress. With the completion of the canal at Lockhart's shoals in 1824, water navigation on the Broad River was possible up to the state line. What was missing was the central link to connect the Saluda and Broad river navigations with the Congaree and Santee river navigations: the Columbia Canal.

The Columbia Canal was to be a short canal beginning in the Broad River just above the dam then under construction. The canal would extend three miles downstream to the Congaree River above Granby and the steamboat docks. Four locks would carry boats down from the Broad River to the Congaree. A basin was to be constructed nearly opposite the State House in Columbia for canal boats and towboats (also called drag boats) from the steamboats at Granby. These towboats would be towed up the Columbia Canal for offloading and reloading.

Construction on the Columbia Canal began in March 1820. The contractor was McKenzie, the same contractor then building the dam across the Broad River. By the end of 1820 one mile of the three-mile canal prism had been excavated.[107] In 1821 the Board of Public Works was delayed in contracting for the construction of the four locks at the lower end of the canal by a provision of the state legislature that required work to be completed in the same year it was contracted for.[108] As a result the excavation of the canal was completed by 1821, but the four locks were not. Until these locks were finished, cargoes from canal boats in the Columbia Canal had to be transferred across a portage to boats in the Congaree River: "No contract however has been made in consequence of the act of the Legislature of the last session, restraining the Board from making contracts which will extend into the next year. This will occasion some delay, as the most favorable season for working will be past, before any contractor can make preparations to enter on such an undertaking. This is much to be regretted, for as long as the Locks remain unfinished, boats from Broad and Saluda rivers, cannot pass from the canal into the river at Granby. They must there transfer their loads by a portage of about sixty yards."[109]

Floods in September 1821 and throughout the winter of 1821–22 delayed construction of the required locks. The recurring sickness that plagued canal construction also delayed work. By the end of 1822, the Board of Public Works was reporting: "The canalling here [on the Columbia Canal] has been finished and the work answers the best expectations of the Board. The banks stand well and have received but little injury, although the season has been unusually rainy. The dam and locks are not finished. The causes of this delay it is necessary to state. During the whole of the last winter and spring, there was a continued succession of freshes in the Broad river.... [Then] the sickly season commenced."[110]

The board reported on the status of the uncompleted locks: "The locks at Granby were to be built under two contracts. The three upper ones were to be of brick and the lower or river lock of stone. Two of the brick locks will probably be up by the 1st of December [1822], and the other may be finished by January [1823]. The stone for the river lock is prepared and the pit dug out. The contractors having failed, this work is suspended, when resumed it may be completed in three months."[111]

By 1823 all work on the locks on the Columbia Canal had been completed except the river lock. Blanding, the new superintendent of public works, had difficulty obtaining a contractor to complete this lock:

> According to the expectations of the Legislature, the work was required to be finished by December. The previous preparations for the work would nearly consume all the time before [the onset of] the sickly season. The site was regarded as unhealthy, and was particularly exposed to freshes which have been known to rise more than 30 feet above the bottom of the lock pit within the last three years. All these circumstances combined to deter bidders. By making a very liberal allowance for delays from water, and for keeping dry the lock pit, I obtained a contract in February, and the dry weather, since the return of the healthy season, has enabled the contractor to get his work so forward, that there is no difficulty in finishing it before the 20th December [1823].[112]

After further delay the lock was completed in February 1824, not December 1823 as promised by Blanding.[113] The canal finally opened the same month. From the Broad or Saluda river navigation, boats with a four-foot draft could now transit the seventy- by sixteen-foot lock chambers into the Congaree River below Columbia. Towboats could also be brought up from Granby and into the Columbia basin. Blanding described the Columbia Canal for the state legislature:

> This canal, which is 3 miles long, was completed in 1822, except the lockage, but was of no use till filled with water by the Broad river dam. Its entrance from above is now perfect.—Much of it runs over difficult ground. There has been much blasting in the upper, and considerable embankment in the lower part of it. There is on it 5 capacious basins; 1 brick bridge, and one wooden one. Another will be required, and the brick abutments to it are built. Whenever the owners of the land, for whose use it is to be erected, requires it, it must be finished. Three waste wiers [*sic*], and five large stone culverts are on it. It has a strong guard lock finished, and four lifting locks, three of brick are finished, except hanging the gates.[114]

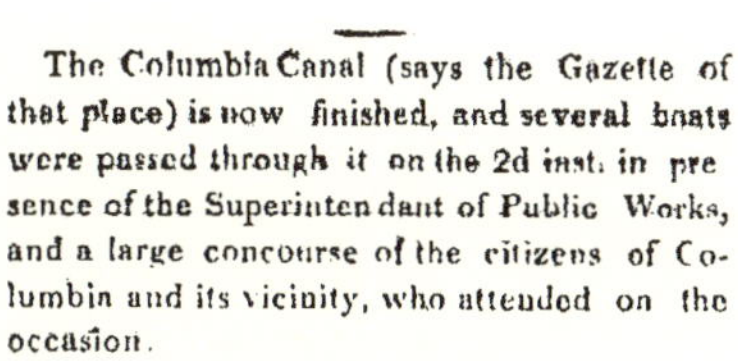

The Columbia Canal (says the Gazette of that place) is now finished, and several boats were passed through it on the 2d inst. in presence of the Superintendant of Public Works, and a large concourse of the citizens of Columbia and its vicinity, who attended on the occasion.

Southern Patriot, February 13, 1824

Individual costs for the Columbia Canal, the Broad River dam, and the Bull Sluice Canal are not known. The Board of Public Works reported the cost of all three structures as $203,187.31.[115] The Columbia Canal became the most widely used of all canals constructed by the state.

Improving the Wateree and Catawba River Navigations: Wateree Canal, Landsford Canal, and Rocky Mount Canal, 1818–1828

Work was also begun on the Wateree River, which meets the Congaree southeast of Columbia in the vicinity of Fort Motte, northeast of present-day St. Matthews. Opening the lower Wateree would open steamboat and canal-boat travel between Camden and Charleston. Furthermore improvement of the Wateree River and the Catawba above it would open the central portion of South Carolina to commerce with Charleston.

While directing the improvements of the Congaree and Santee river navigations, Wilson also began improving the navigation of the Wateree. The lower Wateree, similar to the Congaree and the Santee, could be opened to navigation through clearing of sandbars and removing logs and other snags. In 1818 Wilson reported that this work would be relatively easy: "The navigation of the Wateree is impeded by the accumulation of logs, by sand bars, by gravelly shoals, and in some few places by rocks which can be easily removed."[116]

In 1823 Blanding confirmed the favorable nature of the Wateree for navigation: "This stream is narrower and deeper than the Congaree, and is capable of having a better navigation from the Santee to Grave's ford, 8 miles above Camden, where its first falls commence. Its whole length to these falls is about 50 miles."[117]

Work on the river had begun in 1819. In January of that year Wilson awarded a contract to Mordecai Barbour for clearing the Congaree, Santee, and Wateree rivers—and to perform similar work on the Broad River.[118] Wilson's intention was for Barbour to deploy one hundred men and four flat boats for clearing operations on these rivers during 1819:

> The principal object of this contract was to place without delay a large and competent force on the Wateree and Congaree rivers, as two companies had been formed for navigating them with steam boats. A sum of money was placed in the hands of the contractor, to enable him to forward workmen to Columbia, and to furnish himself with such materials and tools as were necessary for the undertaking. In order to commence the works at once which were designated in his contract, a force of one hundred men should have been immediately employed; this number would have given eighty men to the Wateree and Congaree rivers, by working four sets of flats with machinery.[119]

Starting from the Congaree River, Barbour and his men worked north up the Wateree. By the end of the year they were halfway up the river to Camden and stopped work at the river obstruction called the "raft," a mass of trees difficult to remove at Bentley River.[120]

Barbour did not meet Wilson's expectations, and the contractor was dismissed:

"In the month of March he [Barbour] brought with him to Columbia Thirty-three men; and in the first week in May only one set of flats was prepared for the Congaree river.... On the 31st of August, as I had reason to be dissatisfied with the manner which these works had been conducted, the Contractor was directed to suspend the work."[121]

The following year, 1820, Colonel William Nixon was given the contract for the work.[122] Blanding's instructions to Nixon called for him to clear the river using flatboats, employing teams of fifteen slaves per flatboat with each team supervised by a white overseer.[123] Nixon reported his progress to Blanding and the Board of Public Works: "I commenced working on the Wateree river, at Belton's landing, opposite to the town of Camden. The river has been cleared down to Champion's plantation, fourteen miles by land, below this place. The channel is opened one hundred feet wide, the banks trimmed and the trees cut down or girdled, where there was danger of their falling into the stream."[124]

The board was optimistic that Colonel Nixon and his men would open the Wateree River to navigation to Camden that year: "The continual rains, high freshes and an unusually sickly season, have, in some degree, retarded the works on the Santee and Wateree; but we have reason to believe that they will be completed in the course of the next year."[125]

Later the same year Colonel Nixon had a more optimistic report for Blanding:

> My flats are now within three miles, (by water) of the place called silver gut, the most obstructed part now remaining on the river. It is a cut four hundred or five hundred yards in extent, through which the river forced its way fifteen or twenty years ago. The whole of the timber that grew there before, is now in its bottom. If I am not hindered by high water, I expect to be there in ten days and to have it cleared by Christmas [1820]. By the last of January [1821], I hope to have this river half done, from Camden to the raft, up to which Mr. Barbour worked last year, and this much the worst half; so that two setts of flats and thirty-six hands will be sufficient to finish this river in one year.[126]

Nixon explained to Blanding why he had been delayed in completing the work on the Wateree: "Had it not been for the continued rains and high water, it would have been nearly finished by February next [1821]. But during the whole summer [of 1820], I never could have more than two weeks good river at a time, so that I could work on the channel to advantage. The river has been up thirteen times since I commenced in February. Six of them were high freshes. Every time the water was up, I lost ten to fifteen days, during which, work on the channel of the river could not be carried on to advantage."[127]

During high water Nixon kept his men busy building a one mile long road from Camden to the boat landing on the Wateree River. By the end of 1820 this road, although not completed, was passable.[128]

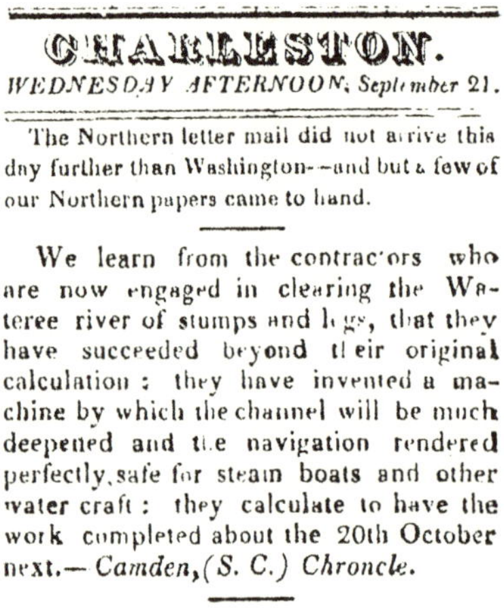

CHARLESTON.

WEDNESDAY AFTERNOON, September 21.

The Northern letter mail did not arrive this day further than Washington--and but a few of our Northern papers came to hand.

We learn from the contractors who are now engaged in clearing the Wateree river of stumps and logs, that they have succeeded beyond their original calculation: they have invented a machine by which the channel will be much deepened and the navigation rendered perfectly safe for steam boats and other water craft: they calculate to have the work completed about the 20th October next.—*Camden, (S. C.) Chroncle.*

Southern Patriot, September 21, 1825

At the end of 1820 the board optimistically reported to the legislature: "Another year will render this river navigable by steamboats, to Camden."[129] Despite this prediction, the Wateree River was not opened to steamboats in 1821. No work was undertaken that year on opening the Wateree River to navigation below the Grave's Ford Falls (eight miles above Camden). In 1822 and 1823 only a small amount of work was done. The ability of steamboats to navigate on the lower Wateree by the end of 1823 was the same as it was at the end of 1821.[130]

In 1824 Blanding awarded a contract to Colonel Nixon and his partner, a Mr. McRa, for eight thousand dollars for clearing most of the remaining obstructions on the Wateree River.[131] Blanding reported: "The contractors commenced their work in April, and aided by the very dry season, made great progress until they were driven from the river by the September fresh; they had cleared out Silver Gut and about half of Bentley river, and a great part of the Wateree from Camden to Cook's mount. Another dry season will be required to complete the work according to contract."[132]

The floods of September 1824 and December 1824 hindered Nixon and McRa's work. The December flood damaged much of the repair work they had already accomplished: "The work was not finished, when the fresh of December rose to great height and demolished much of the work which had been partly finished: so that most of it was required to be reconstructed."[133] It was not until 1826 that the navigation of the lower Wateree River was finally completed.

In 1820, while work was underway on opening the channels in the lower Wateree River, the Board of Public Works began work on the Wateree Canal on the upper Wateree River to bypass Jones' Mills, Loves' Shoals, and Grave's Ford Falls. These obstacles in the river here covered a distance of six miles. The lowest of these three rapids was Grave's Ford Falls. Blanding described these obstructions to the state legislature (proceeding downriver):

1st Jones' falls, in about half a mile, are six feet and are followed by about one mile of deep, still water.

2nd Loves' shoals then begin and in two miles more reach the head of Stark's island, with a fall of twenty-five feet fifty-six hundreds. Between Stark's island and the right bank of the river is a natural canal which receives Sawney's creek, from the west and again enters the river in the distance of about one mile, having six feet fall. From this point to a sheet of still water above Graves', a distance of about one mile, there is a fall of six feet forty hundredths, and from the foot of this still water to the bottom of Graves', there is a fall of eight feet sixty hundredths. Below this point, the Wateree has no natural obstruction of any magnitude.[134]

The board had to build a single canal on the western bank of the Wateree, where the ground was found more favorable. Bypassing all three obstacles, the Wateree Canal was 5.75 miles long with seven locks (later increased to eight) overcoming a fall of fifty-two and a half feet. The southern end of this canal was immediately below Grave's Ford Falls (now located in Lake Wateree).[135]

In 1820 the board entered into a contract with Messrs. Kibbe and Grafton for constructing the upper section of this canal, from Jones's mills to Sawneys Creek.[136] One immediate problem facing the contractors was the lack of stone in the immediate vicinity of the new canal. Transporting stone to the canal would have been prohibitively expensive. Blanding authorized the use of brick to build the locks, except for the hollow quoins, the counterarches, and the breasts on the lock. Brick locks were less expensive than stone, but they were thought to be not as durable.[137]

In his instructions to Walter Izard, his assistant engineer, concerning these locks, Blanding emphasized the importance of having the contractor use good roman cement (hydraulic cement) and good, clean sand to insure that the mortar and cement in the masonry locks would be of the highest quality.[138] By 1821 the board could report: "The excavation of this canal has been extended about two miles through the deepest and most difficult part of the line. All the dressed stone for the coping breasts and heel stones of three Locks, have been prepared, and the bricks for two of them have been moulded."[139]

Blanding, reported in 1822 that four miles of the canal had been excavated and substantially finished.[140] Although work had been suspended that summer because of the death of the contractor and additional work still needed to be done, Blanding believed that all of the necessary work on this section of the canal could be completed by May 1823.[141]

What remained to be done was to build the lower, or second section, of the Wateree Canal, a distance of 1 mile and 60 chains (1 mile and 3,960 feet or 1.75 miles). Blanding recommended that the contract for the second section be held until the first section could be completed "for most of the materials required for it [the second section], must be brought down the canal or be subject to land carriage for about twelve miles over very bad roads."[142]

Even without the Wateree Canal, Blanding thought that the number of boats descending the Catawba River to the Wateree River could be increased by reducing the length of the portage at the Great Falls of the Catawba River. Blanding recommended that the portage be reduced to two and a quarter miles, predicting: "When this is done, even without completing the second section of the Wateree canal, the trade of the country on the upper part of the Catawba extending a considerable distance into North-Carolina, can descend to Camden and Charleston. For although the shoals at Loves and Grave's are difficult, yet when there is a small swell in the river, boats carrying 30 to 40 bales of cotton can pass them; and at the present time most of the cotton raised as high as Rocky Mount, descend the Wateree river in boats."[143]

By 1823 the first section of the Wateree Canal had been completed. It consisted of one guard lock, three bridges, three waste weirs, and eight culverts. The works had "been constructed of brick, secured by stone of the best quality." Four miles long, this section cost $67,692.86.[144]

The second section was more difficult to construct. In 1823 Blanding reported on two alternatives for this lower section: "Two routes present themselves. The first passes Sawney's creek, and requires an aqueduct 80 feet wide to pass the water of that rapid stream under the canal.—Some distance below the creek, the line of canal for near half a mile, runs on the precipitous sides of hills which come to the water's edge, and are composed of hard clay slate, difficult to blast, and too hard to be quarried, without blasting. The other line contemplates entering the river above Sawney's creek, and passing Stark's shoals and Gravesford by two locks in the river."[145]

Blanding also reported that materials would be difficult to acquire for either of these routes: "Stone for the construction of these locks cannot be had, but by hauling it over land four miles, and boating it down the river and canal 10 miles further."[146]

Blanding recommended that the second route be undertaken, at an estimated cost of $76,635.[147] But he reported that "the course to be pursued must be determined by the Legislature."[148] Subsequent reports do not discuss which alternative the legislature chose, but surviving maps indicate a single bypass canal without recourse to the river.

In 1825 Blanding reported that the lower section of the Wateree Canal was nearly completed although "work on this canal has been carried on with some interruption occasioned by the sickness of the workmen in the autumn."[149] He predicted that the canal would be open in January 1826.[150] Charleston newspapers reported in that month:[151]

> **We have been informed, says the Columbia Telescope, by the Superintendant, that the Wateree Canal is now completed, and a considerable quantity of cotton has already descended through it.**

Southern Patriot, January 26, 1826

Above Wateree Creek the Wateree River becomes the Catawba River. Opening the Catawba River to navigation would open a substantial portion of South Carolina and a portion of North Carolina to water-bound trade to and from Charleston (see color plate 12). Despite Blanding's expressed optimism, the portage around the Great Falls of the Catawba was never very widely used, and it became apparent that bypass canals would be needed there. As early as 1818, Wilson explained the situation to the legislature: "The navigation of the Catawba river, above Wateree Creek, is obstructed by rapids and falls, and will require expensive works to render it navigable. The importance of this river to the trade of the interior will warrant any expence the legislature may think proper to bestow upon it. The state of North-Carolina has improved the navigation above the boundary line in the expectation that the works below the line would be carried on with corresponding spirit. Since the acquisition of the Charter of the Catawba Company by this state, the North-Carolina Company have renewed their operations."[152]

Unlike other rivers being improved by the state engineer and later by the Board of Public Works, the navigation of the Catawba River had been originally entrusted by charter to a private organization, the Catawba Company. Although this company was moribund by the 1820s, it had been given exclusive rights to develop the river. Before the state could begin construction of the canals, the legal problems posed by the existence of this company needed to be addressed. The Board of Public Works had to acquire the company's charter. The drawings and engineering data held by the company also had to be acquired.[153] Complicating things further, "most of the lands ceded to the state, by the Catawba Navigation Company, which were granted to them by their charter of 1787, were found to be occupied by persons, who, by a long period of uninterrupted possession, it is believed, have acquired a legal title to them." Rather than initiate a potentially long and expensive legal action to evict these people, the board recommended "to the legislature, the justice and policy of confirming these titles, reserving to the state, the right of taking materials for the public works, and of running roads and canals through the lands."[154]

Besides the legal obstacles to opening the Catawba River to navigation, there were also substantial physical obstacles. The two principal obstructions were "at Rocky Mount, there is a fall of ninety feet in three miles; and at Land's Ford, eighteen feet in less than one mile."[155] These two obstacles were at opposite ends of the Catawba River: Rocky Mount, near the site of the U.S. Military Establishment at Mount Dearborn, was located in the southern portion of the river, and Landsford was located in the northern portion.

At Rocky Mount the board planned three bypass canals and one in-river navigation: the Catawba Canal, the two canals that were called the Rocky Mount Canal, and the use of the Catawba River for navigation (see color plate 12).

In 1820 Blanding explained his plan for bypassing the lower Catawba River obstacles to the legislature:

> From Powell's landing [where the Catawba Canal would begin] to Kingsberry Ferry, the bed of the river presents a rocky and uneven surface, with a very great fall, and in several places, precipitous, which excludes all possibility of rendering the bed of the river a commodious and safe navigation. The only effectual method to overcome the

obstructions, is by a navigable canal [with] Locks and Dams, which may be effected with more ease than might be expected from the adjacent country, which is very broken. But to explain the subject more fully, it is divided in the four sections: First, from Powell's Landing to the mouth of Fishing Creek [later called Catawba Canal]. Second, from Fishing Creek to Gill's old mill [using the Catawba River for navigation to immediately above Mount Dearborn]. Third, from Gill's old Mill to Rocky Creek [west of Mount Dearborn to Rocky Creek, later known as the Rocky Mount Canal]. Fourth, from where the canal enters Rocky creek to Kingsberry's ferry [below Rocky Mount]. The whole distance from Powell's Landing, to Kingsberry's Ferry, is 8 ½ miles, 5 rods and 20 links [8.5 miles plus 95.7 feet or 8.52 miles]. The fall 178 feet, 7 inches and 4 tenths.[156]

Section 1, the Catawba Canal, was 2.36 miles long overcoming a fall in the river of fifty feet.[157] Section 2 was the in-river or slack-water navigation in the Catawba River from the mouth of Fishing Creek to just above the Great Falls of the Catawba, a distance of 1.55 miles. The slack-water navigation was created by dam located just north of the Great Falls of the Catawba.[158] Section 3 was the uppermost part of the Rocky Mount Canal, a bypass canal extending from the river over 2.0 miles due south to Rocky Creek west of Mount Dearborn (see color plates 6 and 7). This section had eleven locks each with a lift of eight feet.[159] The fourth and final section, also a part of the Rocky Mount Canal, connected Rocky Creek to the Catawba River, with three locks and a distance of 2.1 miles.[160] These four sections, taken as a single project, became the most expensive and most time-consuming project undertaken by the Board of Public Works.

The board contracted for section 1, the Catawba Canal, in 1820. Early that year the Charleston newspapers were reporting, "General Davie and the assistant engineer are examining the river, and a strong force of stonemasons, blasters, etc. are expected from the north, to commence the work in April."[161] By the end of the year contractors Thomas and Briggs had excavated one mile of this canal: "A contract has been made with Messrs. Thomas and Briggs, who erected some of the locks on the Schuylkill, for the first section, covering three miles and requiring six locks. The force destined for this work arrived so late that it was thought imprudent to increase it till October. About one mile of the canal is excavated, and some materials for the stone work collected."[162]

By 1821 the contractors had excavated one and a half miles of section 1 and had completed one of the locks: "The Canal has been excavated about one and a half miles and the deep cut of 32 feet between the river and Fishing Creek, is far advanced. The locks are in a great state of forwardness—one has been finished, and another nearly so. Most of the stone for the remaining Locks has been prepared. The materials for this work cannot be surpassed in excellence."[163]

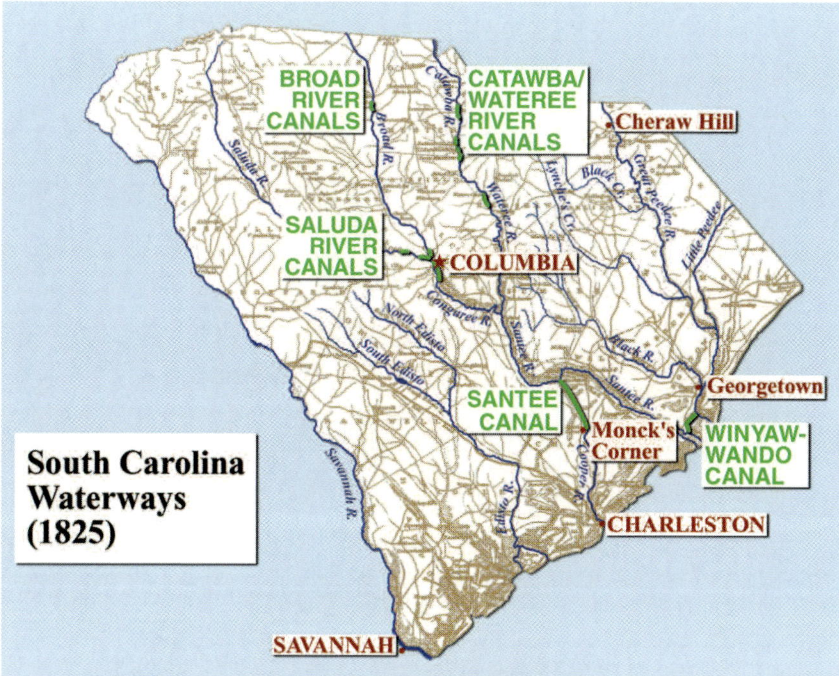

Pl. 1. Waterways and canals of South Carolina. Map by Joan Pennington, Penworks, using Robert Mills's 1825 map as a base

Pl. 2. *City of Charleston Harbor, Looking across the Cooper's River.* Aquatint by William James Bennett, 1838. Courtesy of the Library of Congress, Prints and Photographs Division

Pl. 3. *Right:* The Santee Canal, linking the Santee River at Porchers Embankment with the Cooper River at Biggins Church. Map by Joan Pennington, Penworks, using Robert Mills's 1825 map as a base

Pl. 4. *Below:* Remains of a Santee Canal lock revealed when the drought of 2007–8 drastically lowered the water level in Lake Moultrie. The structures set at right angles to the lock wall are counterforts built to strengthen the wall. Photograph by Jim Huff for the Santee Cooper Corporation, 2008. Courtesy of Santee Cooper

Pl. 5. Biggin Basin at lock 10 on the Santee Canal. Watercolor by Charles Fraser, from *A Charleston Sketchbook, 1796–1806*, plate 29. Courtesy of the Gibbes Museum of Art / Carolina Art Association

Pl. 6. *Right:* "A Survey of the Establishment of the United States at Mount Dearborn" (1809), by Major Alexander Macomb. From *American State Papers, Military Affairs*, 3, no. 362. Rocky Mount is located immediately downriver from this location.

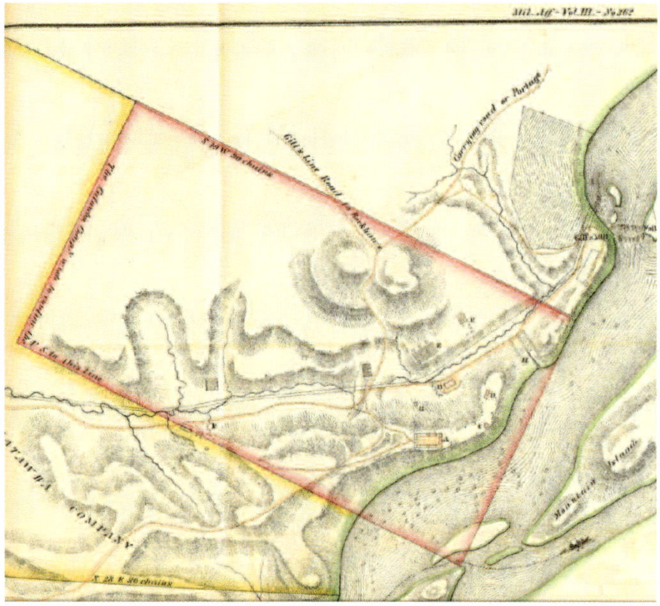

Pl. 7. *Below:* Detail from "A Survey of the Establishment of the United States at Mount Dearborn." The bypass canal began at Gill's Mill in the upper right.

KEY
- A—quarters and barracks
- B—arsenal
- C—powder magazine
- D—artillery shed
- E—huts for laborers
- F—brick kiln
- G—blacksmith shop
- H—site of the proposed armory
- I—tumbling dam (above Gill's Mill)

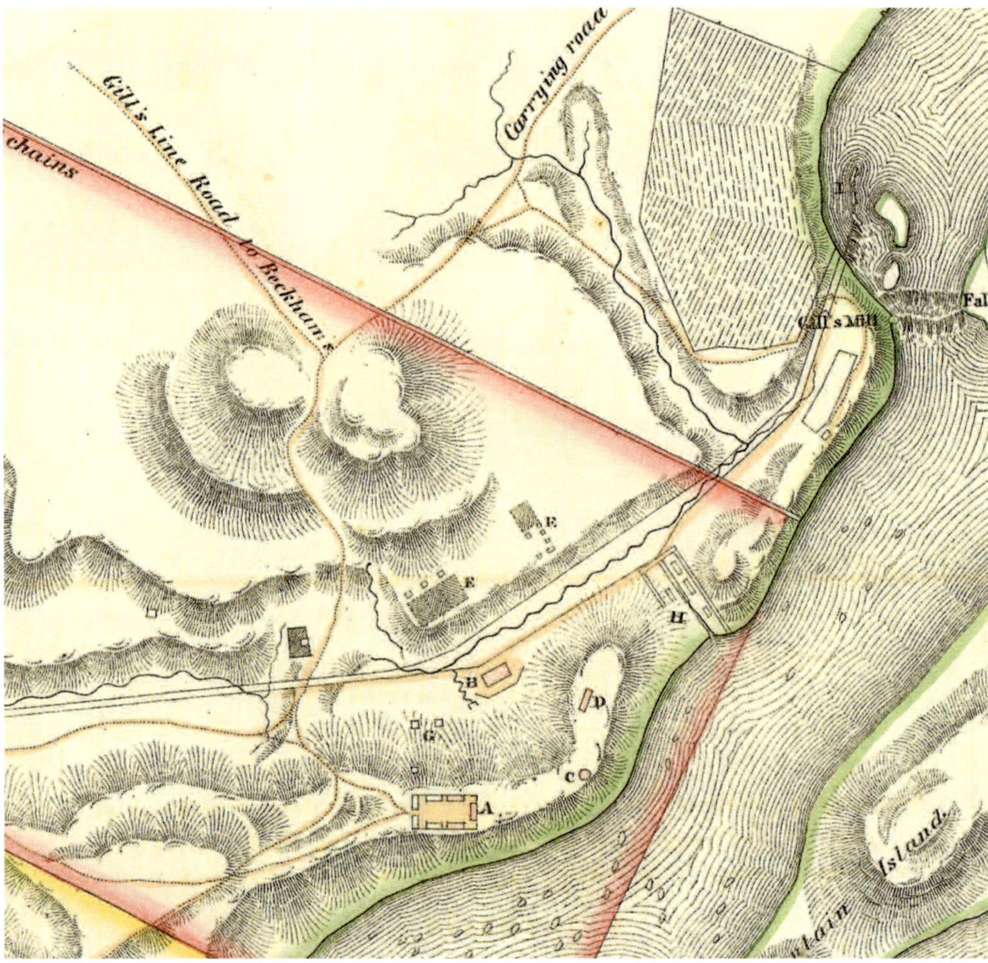

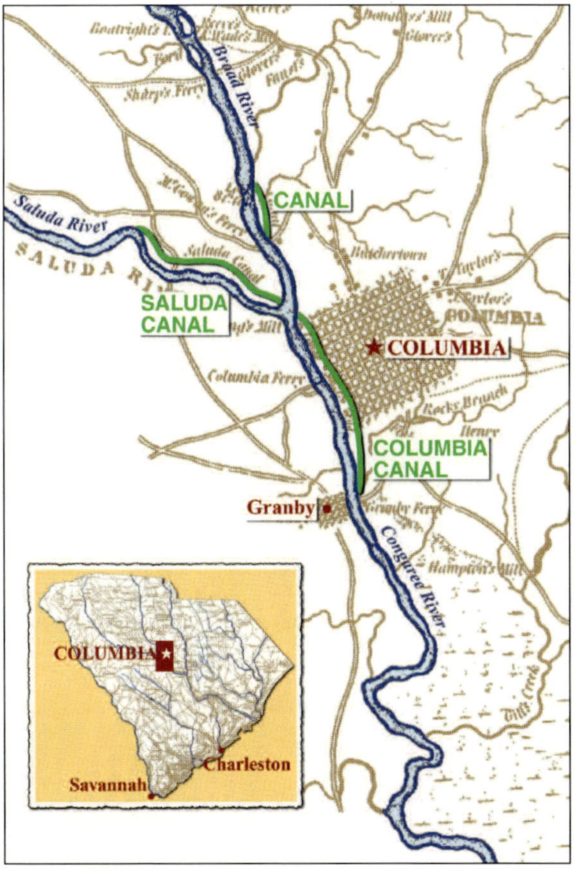

Pl. 8. Canals of the lower Broad River and the Congaree River. Map by Joan Pennington, Penworks, using Robert Mills's 1825 map as a base

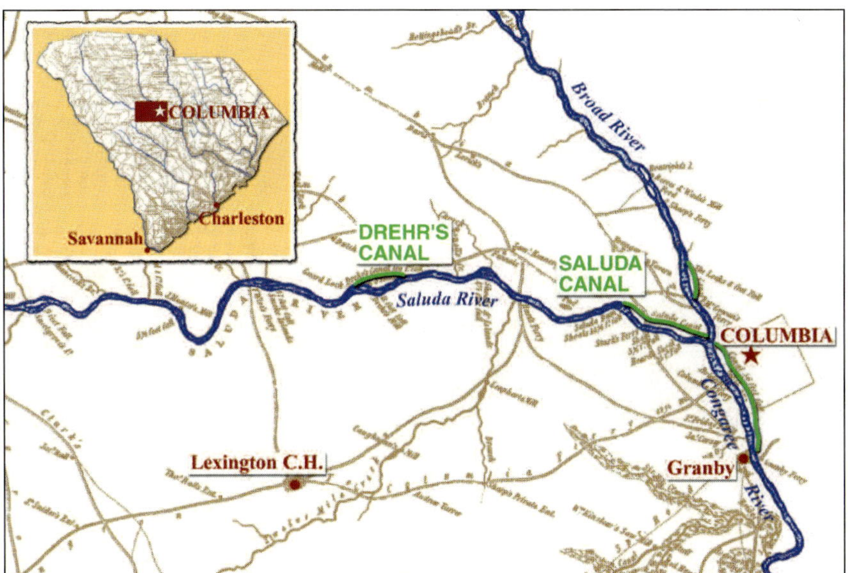

Pl. 9. Canals of the Saluda River. Map by Joan Pennington, Penworks, using Robert Mills's 1825 map as a base

Pl. 10. The Broad, Saluda, and Congaree rivers and the Columbia Canal at Columbia. Detail from an 1872 bird's-eye view of Columbia. By the time this illustration was prepared, navigation had been abandoned on the Broad and Saluda rivers.

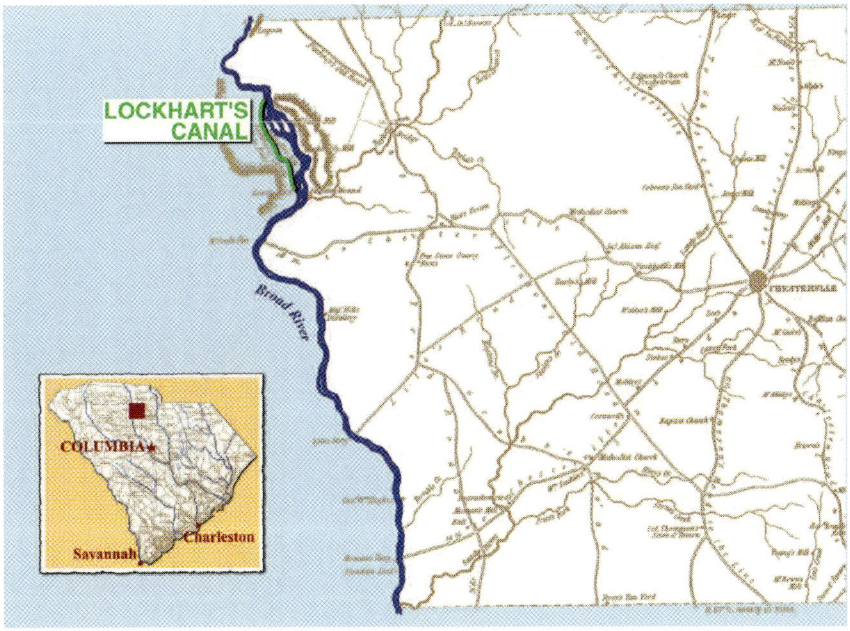

Pl. 11. Canals of the upper Broad River. Map by Joan Pennington, Penworks, using Robert Mills's 1825 map as a base

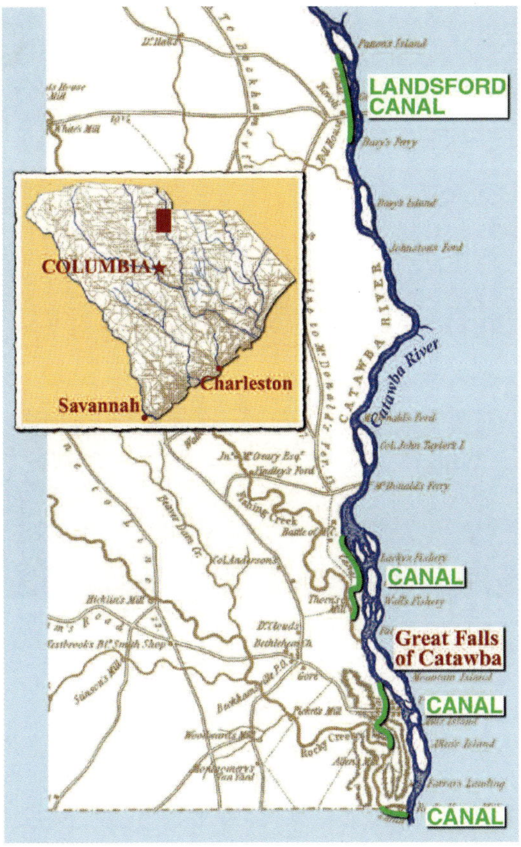

Pl. 12. Canals of the lower Catawba River and the Landsford Canal. Map by Joan Pennington, Penworks, using Robert Mills's 1825 map as a base

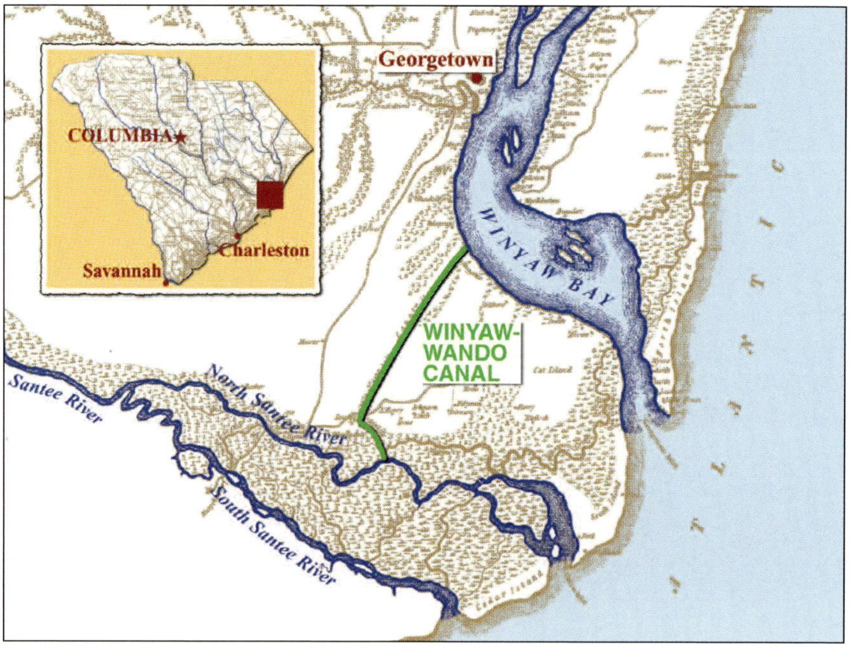

Pl. 13. Winyaw and Wando Canal. Map by Joan Pennington, Penworks, using Robert Mills's 1825 map as a base

Pl. 14.Fig. 5.1. *North-Western View of Charleston,* lithograph by W. Keenan, 1835–55. In the foreground is a typical South Carolina steamboat, built narrow with a shallow draft so that it could navigate the state's rivers. Courtesy of the Library of Congress

By the end of 1822 work on the first section of the Catawba Canal was nearing completion: "There have been constructed on this canal five stone locks of the best materials and superior workmanship. The contractors are entitled to the highest credit for the skill and faithfulness with which they have executed this work. They have got out the materials for the only remaining lift lock, but have not yet done any thing towards the guard lock. All the culverts excepting one, and the side walls are finished, and the whole excavation effected, excepting about eight chains [528 feet]. The work, with a moderate force, may be finished by June next [1823]."[164]

At this point the legislature had not yet decided to build the remaining three sections. Work on the remaining sections did not begin until 1826. In the meantime boats from upstream used the portage road to transport their cargoes around the Great Falls.[165]

In 1826 Blanding awarded a contract to Charles McCullock for section 2, the slack-water navigation down the Catawba River, and for section 3, the excavation of the Rocky Mount Canal to be built west of Mount Dearborn to Rocky Creek:

> With Mr. Charles M'Cullock I contracted for constructing three stone locks, each 8 feet 8 inches lift, on the second section of the Rocky Mount canal, and to cut the canal through the third section to the banks of Rocky creek for $37,650. This contract also embraces the construction of one bridge and a waste wier [*sic*] on the 1st section, which were found necessary, and a road from the hill west of Mount Dearborn to Farrows landing. Mr. M'Cullock has permission to use the upper section of this canal for the purpose of boating down his stone, and he has engaged to keep it in repair without charge. The Work embraced in this contract has been carried on by the Contractor under the direction of Mr. Izard, with energy and success—and I have full confidence that it will be completed by the middle of next year, in a style highly creditable to the undertaker and the Engineer.[166]

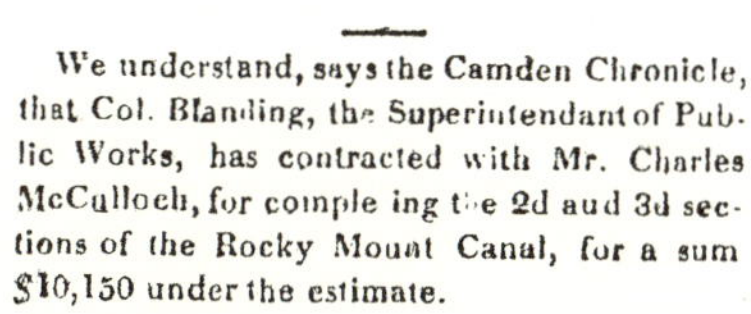
We understand, says the Camden Chronicle, that Col. Blanding, the Superintendant of Public Works, has contracted with Mr. Charles McCulloch, for comple ing the 2d and 3d sections of the Rocky Mount Canal, for a sum $10,150 under the estimate.

Southern Patriot, January 25, 1826

The Rocky Mount Canal was not completed until 1828.

While contractors began to work on the Catawba Canal, the board began plans for a bypass canal to overcome the most northernmost obstacle to navigation on the Catawba River, the rapids at Landsford. Once the Landsford, Catawba, and the Rocky Mount canals were completed, boats from North Carolina could descend three hundred miles to Charleston: "The completion of these works will open the navigation to the North Carolina line. Relying on the good faith of this state, the Catawba Navigation company in North Carolina, have commenced very extensive works upon that river, when they are completed, boats will ascend to the foot of the Blue ridge, within fifty miles of the navigable waters of Tennessee, affording a water communication through a fertile country for more than three hundred miles, and a direct channel for the trade of the western states"[167] (see fig. 4.4).

Fig. 4.4. Rocky Mount Canal lockhouse, the last remaining canals-period lockhouse in South Carolina. Photograph by Frederick D. Nichols, 1930s. Historic American Buildings Survey, Prints and Photographs Division, courtesy of the Library of Congress

The Landsford Canal was located in the upper reaches of the Catawba River, fifteen miles west of the town of Lancaster, and close to the North Carolina border. It was to be 1 mile and 5,016 feet long (1.95 miles) and would consist of one guard lock and four lift locks to overcome a 30-foot fall in the river. Other structures on the canal included six culverts, two waste weirs, two wooden bridges with stone abutments, and a lockkeeper's house.[168]

In 1820 Blanding reported his plans for the Landsford Canal to the legislature:

> The whole fall in this part of the Catawba river, is thirty feet, in the distance of one hundred and forty chains [9,240 feet or 1.75 miles]. At a point above forty chains [2,640 feet or half a mile] below the head of the falls, a race has been cut about thirty chains [1,980 feet or 0.38 miles], which supplied the mills of the late General Davie. It is of sufficient width for a good navigation.
>
> ... By taking a canal out of the river, at the head of the falls, which is four feet above the water in the mill race, entering that race after raising the water in it four feet, continuing that level twenty-five chains [1,650 feet or 0.31 miles] below the mill, descending by a double lock to a new level, which extended to the river, will enter it by another double lock.[169]

In that year, 1820, the board awarded the contract for the Landsford Canal to Robert Leckie,[170] a Scottish mason most recently from Maryland.[171]

By 1821 Leckie and his men had made good progress on the canal: "The Board are highly satisfied with the progress and execution of this work. The excavation has been nearly completed. Two locks, two bridges and the heavy parts of the stone wall have been executed. Materials to a considerable extent are got out for the other locks. The foundations of which are preparing."[172]

By 1822 the board was predicting that Leckie would complete the canal by May 1823: "The canal has been excavated, except for a distance of about ten chains [660 feet], and the four lifting locks, which are all that are required at these falls, have been built. The stone for the guard lock has been prepared. There will be no difficulty in any state of the river to complete this work, and put it into operation by the first day of May next."[173] The Landsford Canal was not finished in May 1823 but rather by autumn of that year,[174] at a total cost of $122,900.13½.[175]

Unlike those on the Wateree Canal, the structures of the Landsford Canal successfully survived the waters of the September 1824 flood. However, in the previous month, the forebay of lock 2 collapsed. Investigation showed that the lock had been constructed on quicksand—sand permeated with springs that gradually eroded the foundation. Blanding had Leckie repair the foundation by installing timber pilings upon which to rebuild the failed lock.[176] Leckie and his men completed the repairs in 1825.[177]

In 1824, before the end of the project, Leckie brought charges against John Couty, assistant engineer to Blanding. Leckie charged that Couty had tried to bribe him to hire Couty's brother, James Couty. In a letter to Couty dated November 12, 1824, Blanding explained the nature of Leckie's charges:

In a conversation I had with Mr. Leckie in August last [1824], and in a further conversation last month, Mr. Leckie stated, that while he was engaged on the Landsford canal, you made a proposition to him, to give your brother James Couty, the contract for the lock gates of that canal. That he refused to do so at the price he was to receive for them and required a profit to himself that to induce him to give your brother the same price Mr. Leckie was to receive, you proposed to make up the difference by allowing him an over measurement of his work. That he rejected the proposition with indignation and would have communicated it to the Board of Public Works, but that he was advised by Mr. Hyder A. Davie not to do so. Mr. Leckie has presented me with Mr. Davie's certificate to the following effect.[178]

Davie's certification of these charges against Couty read:

I hereby certify that Robert Leckie contractor for the Landsford canal waited on me before the completion of the Locks and requested my advice concerning a proposition made to him by John Couty to the following effect. That he John Couty wished him Robert Leckie to give his brother James Couty the contract for making the gates to the locks and that he Robert Leckie refused, stating that it was the most profitable part of the job. Upon which he John Couty said, he would make it up to him in the measurement of the work. My advice was that he should not say anything but watch him and not report to the Board of Public Works, until he was further into Mr. Couty's intentions, as the internal improvement bill had enemies enough. Mr. Leckie said it was his intention to inform on him immediately before convening with me.[179]

To the above, Blanding added an excerpt from a letter written to him by Leckie on November 8, 1824: "Mr. McCulloch says that John Couty told him (and he is willing to make oath to the fact) that had it not been for the liberal measurement that he (Couty) gave the late Mr. McKinsie, in the contract for the Columbia Canal and for the Bull Sluice Lock, that he McKinsie would not come out with the job. The English of this is, that if he had not over measured the work, Mr. McKinsie would have been a bankrupt and let it be remembered that James Couty was a partner (as far as the wood work went at least) With Mr. McKinsie."[180]

Leckie's charge of November 8, 1824, that Couty had deliberately overestimated the work on the Columbia Canal and the Bull Sluice Canal, escalated matters. It was no longer a single isolated incident. Instead, according to Leckie, Couty was following a more systematic practice of over measurement, affecting some of the largest projects undertaken by the superintendent of public works.

Blanding took a cautious approach with regard to these charges: "These are the only direct charges made against you by Mr. Leckie and to which I wait your answer as soon as you can conveniently prepare it. In the mean time, suspending my opinion in relation to them, it is proper that I would wait your vindication before any business is assigned you in this department."[181]

On November 20, 1824, Couty responded to Blanding: "And first I do declare that the direct charges made against me by Robert Leckie, as detailed in your letter to me—both the one stated in the certificate of his former communication to Mr. Hyder A. Davie, and the other respecting the over-measurement of Mr. McKenzie's contract for the Columbia canal & Bull sluice Lock are utterly false; and I never directly or indirectly made to Leckie or to any other person any such proposition as is stated in that certificate to have been alleged by him, nor did I ever over measure or say I had over measured or would over measure the contract of Mr. McKenzie."[182]

With regard to Leckie's allegation that Couty tried to force Leckie to hire his brother and by promising to overestimate the work on the locks to compensate Leckie for doing so, Couty responded that Leckie had hired his brother for this work three years earlier and asked, if Leckie's allegation were true, why did Leckie wait three years to report it? Couty suggested that these allegations began only after he made a measurement of the work completed by Leckie that Leckie disagreed with:

Robert Leckie in the fall of 1820 was engaged by the Board of Public Works to execute the Landsford Locks and Canal. In the summer of 1821, probably in the month of June, he made a subcontract with my brother James Couty to build the lock gates. In this contract I had no interest and with the exception of my own immediate engagement I have never had any interest in any of the contracts either directly or indirectly made by the Board of Public Works since I have had the honor of being employed by that Board. In Octr. 1822, in order to enable me to make my report to the Board previous to the meeting of the Legislature, I measured the work that had been done at the Landsford Locks and canal. The measurement was made according to the usual way and my best judgment, with perfect equity and impartiality. Leckie, afterwards understood, complained of it and expressed himself in terms of much dissatisfaction. A further admeasurement was made in March & in April 1823 sometime before the completion of the work, and again in the fall of 1823 when it was finally completed. At one of the admeasurements, I do not remember which, some of the members of the Legislature were present at a dispute which took place between Leckie & myself in which he objected to my measurement. Now let it be observed, his contract with my Brother was about June 1821. Had I made him the proposition which he so falsely charges it must of course from the nature of the case, and the terms in which he now states it, have been before this contract. When did he consult Mr. Davie? I should wish that time to be fixed with the greatest possible precision. And you, sir, will see the importance of having this done."[183]

With regard to Leckie's accusations that Couty overestimated the Columbia Canal and Bull Sluice Canal, Couty stated that Blanding had possession of Couty's original estimate and that the two structures still stood. Therefore it would be an easy matter to compare Couty's original estimate against the standing structures to verify if Leckie's accusations were correct: "The works done by Mr. McKinsie [the Columbia Canal and Bull Sluice Canal] remain now as they were when I measured them. My measurement of them & the computations founded on that measurement are in your hands. Let them be remeasured by any competent measurer & if the measurement given in to you by me exceeds a fair liberal measurement such as is usually made of similar works, than I shall admit that I have betrayed the trust reposed in me."[184]

Despite Leckie's accusations, Couty was not removed from his position. It can be assumed that Blanding, on behalf of the Board of Public Works, agreed with Couty. Leckie continued to be difficult.[185] Disputes between Couty and Leckie continued.[186] Probably in 1825, Leckie filed a petition with the state legislature for $2,748.65 additional payment.[187] This was approved by the legislature, but payment was never made.[188] Shortly thereafter Leckie relocated to Washington, D.C., to work for the Chesapeake and Ohio Canal Company.

Great Pee Dee River

A part of South Carolina's canal-construction and waterway-improvement plan was to open the upper portion of the Great Pee Dee River to navigation and thus connect central North Carolina with Charleston via the Great Pee Dee and the Yadkin River, the North Carolina portion of the Great Pee Dee.

The Great Pee Dee River was an important South Carolina waterway. Robert Mills's *Statistics of South Carolina* (1826) provides a nineteenth-century view of its navigability: "The Pedee river rises in North Carolina, (where it is called the Yadkin,) and enters Winyah bay above Georgetown, to Cheraw, above 120 miles from the ocean; it has a good steam-boat navigation; from Cheraw, to the North Carolina line, there is a fall on a regular inclination plane of eighteen feet, and above that line the rapids extend to the narrows, about seven miles by water, where the fall is very great. Above the narrows to the mountains, this river is represented as favorable for small boat navigation. It heads near New river, one of the main branches of the Great Kenhawa."[189]

In 1818 Wilson provided the state legislature with an overview of obstacles to navigation on the Great Pee Dee River:

> The Peedee river is navigable for boats drawing four feet water, from Winyaw bay to Long Bluff.[190] Some logs and driftwood have accumulated and form obstructions which can easily be removed. For ten miles higher up, the channel is no where less than thirty feet wide, and two and a half deep at low water.
>
> The obstructions to the navigation between this point and Cheraw Hill, are three rapids, where the water falls over loose stones and rocks, and at times is not more than two feet deep.
>
> Above Cheraw Hill, to the North Carolina line, (less than ten miles in a direct line) there are some ledges of rock, which extend across the river, and the navigation through the distance is obstructed by loose rocks. The water here is represented never to fall below two feet.[191]

It was thought that improvement to navigation on the Great Pee Dee River would open central North Carolina to commerce with South Carolina via the Yadkin River,[192] permitting North Carolina's produce to be transported to South Carolina ports. North Carolina had recently chartered the Yadkin Navigation Company, and by 1818 the company had surveyed the Yadkin from Wilkes Court House in the western part of North Carolina to the mouth of Westfields Creek, above Cheraw (also called Chatham)—a distance of 247½ miles. The company estimated that $250,234 would be required to open this river to navigation by boats of ten tons. In the same year, the company reported that it had stock subscriptions of $170,000.[193]

It appeared that there was some urgency in opening the Great Pee Dee for navigation. The Yadkin Navigation Company had announced that it intended to seek a charter with the South Carolina legislature for exclusive rights for navigation of the Great Pee Dee River within South Carolina.[194] If their plans were approved by the South Carolina legislature, it was thought they would damage South Carolina interests. In February 1819 newspapers carried an announcement that the Yadkin Navigation Company had contracted to open the Yadkin River for navigation.

YADKIN NAVIGATION COMPANY.

Extract of a letter from one of the gentlemen, composing the above Company, to the gentleman in this place appointed to receive subscriptions for Shares, dated

"SNEEDSBOROUGH, 26th Oct 1818.

"Since August, we have had the river examined and surveyed, from Wilkes court-house, in the mountains, to the mouth of West field's Creek, above Cheraw, and find the distance two hundred and forty-seven and an half miles.—The expense of making it navigable for boats of ten tons, is estimated at $250,234, exclusive of the narrows, where for the present, we intended to make a turnpike road; distance about 7 miles; expence about 2000 per mile. $63,400 of the above sum is required to be expended between Sneedsborough and Cheraw Hill; distance 6 1-2 miles, the whole distance, except 200 yards in South Carolina. The amount of our stock subscribed, is about $170,000. We again opened books, which will continue open until the 1st Thursday in January. We hope, in that time, to make up $250,000. We intend to increase our capital to whatever sum may be wanted, and hope to obtain something from the Legislatures of North and South-Carolina, to whom application will be made at their next Sessions.

(Article continues on next page)

These actions of the Yadkin Navigation Company had motivated Wilson to investigate the Great Pee Dee River for navigation. In 1819 he reported to the legislature that he had directed a survey of the river for that purpose. He instructed his assistant, Captain Sobieski, to "explore the Great Pedee River, and to examine all the obstructions to its navigation from the North-Carolina line to Tide water. To take such levels, and to make such drawings as were necessary to the elucidation of this important object. This Report was furnished on the 20th January last [1819]; it described minutely the nature of deposites which have been formed in the River from that point where the tide ceases to Gordon's fall; and from Whortleberry Creek, a line of levels were designated along the margin of the Pedee for the distance of ten miles, with a view of meeting the Canal already begun by the Yadkin Company of Whortleberry Creek."[195]

Wilson and Sobieski found that the rapids on the upper part of the Great Pee Dee River could be bypassed by means of canals. The problem was that these canals could not be protected from the periodic floods: "Although the margin [of the Great Pee Dee River] affords a fine soil for digging, I cannot recommend to the legislature the prosecution of this work. The whole line of the canal would be exposed to the freshets of the river, and would require substantial and extensive aqueducts."[196]

By 1820 the situation with regard to the North Carolina had changed. It no longer appeared likely that the Yadkin Navigation Company would be able to open the Yadkin River for navigation.[197] Consequently the board decided that efforts should be concentrated on clearing the Great Pee Dee River below Cheraw[198] for steamboat operations rather than opening the river to the state line: "It was thought advisable to direct our operations to remove the obstructions to the navigation of the Pee Dee, below Chatham [Cheraw]. This has been effected under the direction of General D. R. Williams, to whose zeal and activity the state is indebted for the completion of this important work. There are some shoals between Chatham and Society Hill, which impede, although they do not altogether obstruct the progress of the steamboat."[199]

Clearing the Great Pee Dee River of logs and sandbars continued throughout 1819 and 1820 under the direction of General David R. Williams.[200] Blanding instructed General Williams to use six flatboats, each manned by a black coxswain

Gen. Pearson (one of our directors) will attend your Legislature. The object in view, is to obtain from South-Carolina an extension of our charter, so as to take in the whole river to Georgetown. If, under their present plan of internal improvement, that is not admissable, then have the exclusive right to such improvement as they may make by canals, leaving the bed of the river to be improved by the State. The distance from Cheraw to Georgetown by land is 103 miles, by water, it is called 270 miles, at least one third of that distance can be saved by canals at a less expense, than clearing the river of logs. If we cannot obtain that, (and if the State will do it we do not want it) we hope to have granted to us, the right of cutting a canal and looking down from Speedsborough to Cheraw Hill, a distance of 6½ miles. The advantages that will result to South Carolina, will be very great; the trade of near one fourth of our state, will go down the river. The county of Anson made last year, about 5000 bales of cotton; 3500 of which was waggoned to Fayetteville; 1500 went to Charleston, New York and Philadelphia; from which, I think, we may fairly conclude the trade from this state to South-Carolina, will be worth nearly as follows:

Anson County, - - -	$350,000
Richmond, - - - - -	200,000
Montgomery, - - - - -	250,000
Rowan, - - - - - - -	500,000
Cabarrus, - - - - - -	150,000
Iredell, - - - - - - -	150,000
Surry, - - - - - - - -	350,000
Stokes, (in part) - - -	100,000
Wilks, (the Whole) - -	300,000
Ash, - - - - - - - - -	50,000
Burk, (part) - - - - - -	35,000
Randolph, (part) - - -	25,000
Guilford, (part) - - - - -	25,000
	$2,475.000

The whole will not be less than two millions, and perhaps up to this statement, being nearly one fourth of the trade of the State."

☞ A subscription paper for Stock in this Company is continued open for subscribers, at the Georgetown Book-Store.

Winyaw (Geo.) Intelligencer, Nov. 4.

Southern Patriot, November 5, 1818

and ten black laborers. These instructions still exist in the state archives.[201]

In 1820 Williams had assembled a workforce of some fifty-five men under the supervision of John Barnett to clear obstructions in the Great Pee Dee River.[202] By August the Charleston newspapers reported that the river was entirely cleared.[203] The state expended $16,875.16 for these operations on the Great Pee Dee—a small fraction of what it had spent on the Saluda, Broad, Wateree, and Catawba river systems.[204]

☞ An Annual Meeting of the Charleston Stockholders in the Pee Dee Steam Boat Company, will be held TO MORROW, 9th inst. at 5 o'clock, P. M. at No. 23 Vanderhorst's wharf, to elect Officers for the ensuing year, agreeably to the Rules, when a statement of the Company's affairs will be laid before them.

J. COIT, *Agent.*

March 8 2

Southern Patriot, March 9, 1822

The distance from Cheraw to Georgetown was 120 miles overland and almost twice that by way of the circuitous Great Pee Dee. Once that river was cleared, the steamboat journey from Cheraw to Georgetown took only three days and five days ascending.[205]

Although the Great Pee Dee River was now used by one steamboat and one or two team boats,[206] there were still a number of obstructions between Cheraw and Georgetown that would obstruct navigation during dry months or dry years. In his 1823 report to the legislature, Blanding identified fifteen obstacles to navigation on the Great Pee Dee River during low water.[207] He estimated that eight thousand dollars would be adequate to remove all these obstructions.

In 1824 Blanding entered into a contract for ten thousand dollars with General Williams and David Gregg for clearing the Great Pee Dee navigation channel of the low-water obstructions, including the logs and snags from Cheraw to Georgetown, and for deepening the channel at Thompson's and Campbell's Shoals.[208] In closing his report, Blanding made note of a long-standing problem on the navigable rivers of South Carolina, that the draft of the existing steamboats was too great for the cleared South Carolina rivers: "In closing my observations on this river, it is proper to remark, that the steam boats which have usually run on it, draw about 4 feet 6 inches, and some five feet water. This is too great a draft, and it is vain for the state to attempt works which will enable them to run at all seasons of the year. The boats must be accommodated to the river, and not the river to the boats."[209]

Clearing operations continued on the Great Pee Dee River and its tributaries for the next several years. Most of these operations involved removing trees and stumps from the navigable channel and removing sandbars and gravel bars. But Blanding also constructed jetties along the river, explaining, "The sand bars were generally shifting and the channel through them changed or filled up at every fresh. To keep these channels open it becomes necessary to force the river through a narrower space. This was effected by jettees extending from the shore, at some points nearly half across the river. They are formed by driving plank 12×3 inches, and securing their heads by timber 12 inches square. The piles are driven 6 feet into the bottom of the river or as far as a pile driver weighing 1000 lbs. falling 12 feet can drive them. There have been constructed in this river 12 jettees, which exceed 2500 [feet] in length, or an average of about 200 feet each. Two of them are formed of double rows of piles framed together and filled with stone."[210]

Innovations facilitated the clearing of the Great Pee Dee. To cut the navigable channel to the necessary depth of four feet in the sand, gravel, and clay bars in the river, contractor David Gregg invented a dredge. Blanding described this machine:

> At the gravel and clay bars the dredging has been effected by a machine entirely new; the invention of Mr. David Gregg the contractor. Its construction is as follows:—Two flats are placed about 10 or 12 feet a part and secured by timber crossing them at each end. Between them is a water wheel 10 feet in diameter and about 10 feet long. The arms of this wheel extend a little beyond the paddles, and are pointed with iron in the shape of a bull tongue plough. The flats are moored at the head of the shoal and by a chain and capstan can be let down through it and drawn back again. While the flats are passing slowly down, the wheel is lowered into the water till the iron arms strike the bottom, and being set in rapid motion by the water of the river passing between the flats, the gravel and clay are torn up and carried by the wheel and current into the deep water below. In this way a channel is cut out to any depth required.[211]

Above Cheraw the Yadkin Navigation Company was making very slow progress in opening the Yadkin River in North Carolina, having excavated only a canal two miles long around the first falls. Blanding recommended that nothing be done on the upper Pee Dee River until the Yadkin Navigation Company extended their navigation to the narrows of the Yadkin River.[212]

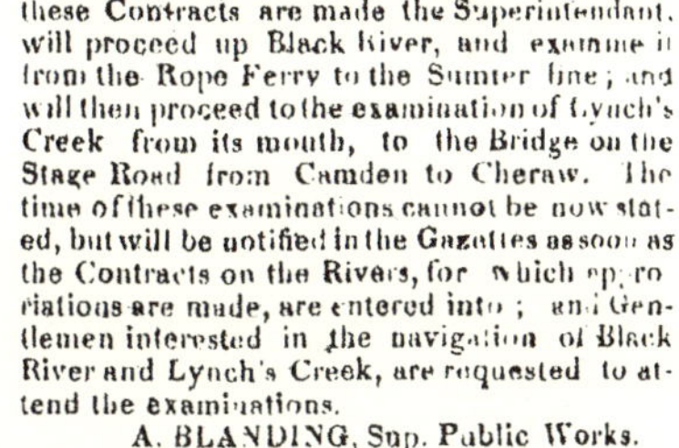
DEPARTMENT OF
PUBLIC WORKS.

THE SUPERINTENDANT of PUBLIC WORKS, gives Notice, that he will ascend the Little Pee Dee, and on Thursday 14th April, at Gullavant's Ferry, will Contract for clearing this stream. On Saturday, the 16th of April, he will attend at Conwayborough, and Contract for clearing the Waccamaw River.—Gentlemen desirous to enter into Contracts, are requested to attend him. Immediately after these Contracts are made the Superintendant, will proceed up Black River, and examine it from the Rope Ferry to the Sumter line; and will then proceed to the examination of Lynch's Creek from its mouth, to the Bridge on the Stage Road from Camden to Cheraw. The time of these examinations cannot be now stated, but will be notified in the Gazettes as soon as the Contracts on the Rivers, for which appropriations are made, are entered into; and Gentlemen interested in the navigation of Black River and Lynch's Creek, are requested to attend the examinations.

A. BLANDING, Sup. Public Works.
April 12

Southern Patriot, April 12, 1825

Below Georgetown work was underway on clearing rivers flowing into Winyah Bay, such as the Waccamaw River under contractor Durant and the Black River and its tributary, the Black Mingo Creek, under contractor Rasha Cannon.[213] The state also cleared the Great Pee Dee River tributaries of Black Creek and Lynches River. In Lynches River a new channel was opened through a series of small lakes, thence to Clark's Creek, and down to the Great Pee Dee River.[214] This work progressed slowly, and by 1827 Blanding was reporting to the legislature that most of the work had not yet been completed.[215]

CAMDEN, (S. C.) May 27.

Col. Blanding, the zealous and indefatigable servant of the public, left this place on Thursday last, for Columbia; having effected contracts for opening the navigation of Black River and Lynches Creek. We regret the exposure and fatigue attendant on his unwearied exertions in the discharge of his public duties, brought on a fever which confined him some days, in this place. We are indebted to his politeness, for the following statement.

The contracts made for the improvement of Black river and Lynches creek embrac a larger portion of those rive than was anticipated. The contracts the first of those streams provide for improvement from tide water to Brewton Lake on the Poco.aligo branch, a to Nelson's Lake on the main branch These points are six or eight miles ab the Sumter district line. When impro the river will admit of the passage boats carrying one hundred bales of c ton, during the winter and spring mon

The contracts on Lynches Creek e brace all that stream from Big Pee D to M'Cullums ferry, near the Kershw line. The improvements when compl ted, will admit of the same navigations on Black River. The clearing of mouth of Lynches Creek, in pursuance of the contract made last year, has progressed so far, that in ordinary states of the water, boats can pass out of the mouth of the Creek When the whole work is completed, boats will be enabled to pass out of the mouth of the Creek, at all times when the state of the water will admit of their passage through the other parts of the Creek. It is highly creditable to Col. B. that he has effected all these contracts considerably within the appropriations.—*Journal.*

Southern Patriot, May 31, 1826

Coastal Navigation: Winyaw and Wando Canal and Other Improvements

Not all improvements to navigation within South Carolina at this time were made by the state. Private companies undertook such projects as well. One example was the Winyaw and Wando Canal, chartered in 1816. In 1819 the company contracted for the construction of a canal between Winyah Bay and the North Santee River, and some progress was made.[216] Completion was expected within a year[217] (see color plate 13).

To finance the canal the company issued twenty shares. Although it did not retain technical direction of the project, the state purchased seven shares. While the project was called the Winyaw and Wando Canal, initially the canal was built only between Winyah Bay and the North Santee River and did not connect to the Wando River, which empties into Charleston Bay.[218]

The canal was constructed in two sections, which Blanding described to the state legislature. The southern section, above the North Santee River, was excavated through a cypress swamp: "From Kinlock's creek, which empties into Santee river, and where the water is first taken out, this canal is excavated through a thick Cyprus swamp; and it may be truly said to be literally cut out of wood, for little else is to be seen upon the banks but chips, roots, and blocks of timber. It extends some distance and exhibits great labor and perseverance. This section was executed under the direction of Col. Thomas Pinckney."[219]

The northern section was excavated through uplands: "The next section is a work of a different nature, being carried mostly through upland, except where it passes through a natural ravine leading into Winyaw bay. This part of the canal was executed under the superintendance of Mr. ____ Hume."[220]

This canal from the North Santee River to Winyah Bay was six miles long and had a width of twenty-five feet with "passing places" added. At each end of the canal a twenty-two-by one-hundred-foot tidewater lock was built of cypress.[221] Started as early as 1819, work on the canal moved slowly because of the difficulties of digging through the cypress swamp. By 1820 only two miles of the six-mile-long canal had been excavated.[222] In 1823 the company began an extension to link the North and South Santee rivers under the direction of contractor Hume. When the Winyaw and Wando Canal was nearing completion in 1823, Blanding reported to the legislature that it had cost the company in excess of $52,000.[223]

The Edisto-Ashley River Canal, 1787–1823, and the Edisto River Navigation, 1818–1820

The Winyaw and Wando Canal helped eliminate the dangers of an ocean trip along the coast north of Charleston. Similar improvements were needed along the coast south of that port city. Attention focused on the Edisto River, south and west of Charleston. A connection between the Edisto and the Ashley River, the Edisto-Ashley Canal, would eliminate an eighty-mile ocean trip—"a circuitous and dangerous route."[224] Improvement of the Edisto River for navigation, much as the Great Pee Dee River had been cleared, would assist in shipping agricultural products downriver and on to Charleston.

The State of South Carolina had chartered the Edisto Canal Company in 1787, but the company had done little to open a canal between the Edisto and Ashley rivers except to conduct several studies of various routes. In 1823 the Edisto-Ashley Canal became the center of attention for the superintendent of public works. Blanding reviewed the results of the previous studies for this canal and reported to the legislature what he thought might be the best route:

> Two routes present themselves. The lower route leaves the Edisto at Hart's cove, about seven miles below Giveham's ferry, and enters the Ashley about half a mile below Bacon's bridge. The distance is 11¾ miles. The fall from Edisto to Ashley is 6 feet 2 inches. The high lands to be crossed, are in the first five miles from Edisto, and the average elevation in this distance is about 7 feet: the greatest 11 feet 7½ inches. The rest of the line passes over ground as low as the surface of the water in the river at Hart's cove. The ground is represented as favorable for canalling and the permanent construction of the banks. Were the waters of the Edisto raised four feet by a dam, the excavation of the first five miles, would average about 6 feet deep, giving the canal 3 feet of water. The whole lockage on this canal, would then be 10 feet 2 inches, and a guard lock to exclude the freshes of the Edisto. The upper route leaves the Edisto at old river, about three miles below Giveham's ferry, and enters

the Ashley at the same point as the first. Distance 13½ miles. Fall from Edisto to Ashley, 12 feet.[225]

Both these routes would have been summit-level canals—canals that surmount a ridge between two watersheds much as the Santee Canal did. Blanding explained: "The intermediate region of the two rivers affords no streams to supply a summit level."[226] The water needed at the summit would have to come from the Edisto River, which was six feet above the Ashley on Blanding's route 1 and twelve feet above the Ashley on Blanding's route 2. The summit level was ten feet eleven inches above the Edisto River, so it was thought that a dam of perhaps four feet in height coupled with an excavation averaging six feet in depth would be required to provide three feet of water for the canal's summit level.[227]

Substantial cost overruns were experienced on most of the state's projects, and by 1823 it was apparent that additional funds would be required to pay for the internal improvements already authorized by the state legislature. The legislature was in no mood to authorize new projects, such as the Edisto-Ashley Canal. Rather than recommending the state undertake the construction of the Edisto-Ashley Canal, Blanding recommended that the merchants of Charleston pay for it. In his report he pointed out that the Edisto River could be a source of water and water power for Charleston as well as being a transportation canal:

> The costs attending this part of the work, its advantages to the health, comfort, and security of the city, and the profits to the company likely to be derived from this source: What extent of rice lands now incapable of profitable cultivation might be watered by this canal, and the probable revenue to be derived from this source: What power may be obtained, and how usefully applied by taking water from this canal?
>
> A careful examination into all these points, may end in results so satisfactory as to induce our capitalists to engage in the enterprize.[228]

The legislature replied to Blanding's report by passing a resolution requiring that "the Superintendent of Public Works to report on the expediency and practicability of opening a canal between the Edisto and Ashley rivers, and on the many advantages that may be derived from the same."[229]

In his response to the legislature, Blanding asked three sets of questions:

> 1st. Can a navigable canal be extended from the Edisto to Charleston? What will be the probable costs of its construction? What its advantages?
>
> 2nd. Can it be made a canal of irrigation? To what extent? To what advantages?
>
> 3rd. Can the City of Charleston be supplied with fresh water by means of this canal? At what cost? With what benefit?[230]

In answering these questions, Blanding came up with two cost estimates: $221,500 for a canal that would provide navigation and water for the City of Charleston and

$127,125 for one that would provide navigation only.[231] This was a lot of money. And, by the time of this report, 1825, enthusiasm for canals had begun to wane. Road improvements were coming into fashion, and the Edisto-Ashley River Canal was not funded.

Unlike the Edisto-Ashley River Canal, the Edisto River navigation did receive support from the state legislature. Like the Great Pee Dee River, the Edisto was seen as requiring only clearing to make it navigable. Robert Mills provided a contemporary description of the river: "The Edisto discharges into the ocean by two mouths, called north and south Edisto inlets. It rises in the region of sand hills in two branches, below Orangeburg; both branches and the main river are navigable."[232] Work on clearing the lower Edisto for navigation began in February 1819. Wilson described the work to the state legislature:

> In the month of February last, arrangements were made for the purpose of commencing without delay, to open the navigation of the Edisto River, and to remove such obstructions as had accumulated in that river, I was aided in this undertaking by Col. John Johnson, Jr., who has shewn great zeal in forwarding the views of the state in relation to this river. Boats with proper machinery for the purpose, and labourers, proceeded from Charleston about the 16th February. They ascended the river, and commenced their operations at Parker's Ferry in March. The distance of this point of the river from the ocean, is about forty miles, and here the influence of the tide ceases.
>
> The accumulation of logs in the river above the Ferry, had been great, and their removal was attended with considerable difficulty.
>
> To enable us to continue the operations on this river, it was found necessary to alter the draughts of the vessels, or flats, by cutting them down in order to lighten them. One large flat and a small one, each fitted with a screw, wheel and chain, have answered exceedingly well.
>
> Notwithstanding the remarkably low state of the river, and the sickliness of the season, the Edisto river has been cleared and made navigable from Parker's Ferry, to the neighbourhood of Givham's Ferry, a distance of nearly thirty-nine miles. This work still goes on. The average number of hands employed, have been sixteen, and the cost, which includes the whole, outfit of boats and machinery, up to November 1st, has been $7698.49.[233]

In addition to opening the lower Edisto to navigation, the state undertook separate projects to open the North Branch and South Branch of the Edisto. Wilson reported in 1818: "The North Branch of the Edisto is navigable for boats and rafts to Black Creek, one of its principal sources, forty-five miles from the confluence. The obstructions which impede its navigation, are sand bars and logs. The general depth of this portion of the North Branch is eight feet. The navigation may be opened to Lightwood Creek, fifteen miles higher up."[234] Wilson also reported on the South Branch: "The South Branch is navigable for rafts and boats drawing five feet water, without any material obstruction, for twenty miles in a direct line, from its confluence with the North Branch. From thence to Goodland swamp, it passes through a low broken swamp, without any channel, and from Goodland swamp, to Dean Swamp, the river is divided into several small streams, interspersed with islands."[235]

While Colonel Johnson was working on the lower Edisto River, a workforce under contractor David Williamson labored on clearing the North Branch, and a separate workforce under contractor James Head cleared the South Branch. Blanding provided detailed instructions to Head on this part of the project.[236]

The work moved quickly, and by 1820 the Board of Public Works was able to report: "The works on the Edisto are nearly finished, and the south branch of that fine stream is navigable to the mouth of Mr. Tyre's creek, sixty miles above the fork for boats ten feet wide and fifty feet long, with a draught of two and a half feet during the driest season. Boats of larger dimensions, say twelve feet wide and seventy feet long, may navigate the river to Head's bluff near the mouth of Shaw's creek."[237]

The Savannah River

Arguably the most important river serving South Carolina, the Savannah River was one of the few rivers not improved by the state. Robert Mills pointed out the potential of the Savannah:

> The Savannah river divides this state from Georgia. It has a ship navigation eighteen miles from the ocean to the city of Savannah, and a good steam-boat navigation 140 miles further, to Hamburg and Augusta. Above these places 100 miles, to Andersonville, the river has thirty-three miles of rapids, with a fall of about eight feet in the mile, on a regular inclined plane; the other sixty-seven miles is a smooth deep water. Boats descend from Andersonville with seventy bales of cotton or ten tons. The Tugaloo is navigable for similar boats twenty-five miles, to Pulaski, and the Seneca twenty-six miles, or six miles above Pendleton court-house. At the junction of twelve mile creek, the Seneca changes its name to Keowee, which is capable of being made navigable entirely within the mountains, by merely sluicing.[238]

For Mills improving navigation on the Savannah River was the means of uniting the western states with the coastal cities by a canal: "The Tugaloo branch of the Savannah rises in the mountains, a short distance from the Hivassee, a navigable branch of the Tennessee river. By means of these streams, it is believed the Southern Atlantic may be connected with the western states, by a navigable canal."[239]

Editorials appeared in the Charleston papers espousing similar views: "The most important stream in the state appears to be neglected, merely because it separates us from one of our sister states. The constituent branches of that stream, lined with the best lands in the state, are capable, with skilful and enterprizing exertions, of the highest improvement in navigation. At their confluence is situated the town of Andersonville, which the enterprize of individuals is now laboring to improve. There can be a good land carriage communicating with West Tennessee, which would save a circuitous rout[e] of fifty miles, and which would arrest the commerce of that territory in its progress to that state, of which we appear so idly and foolishly jealous."[240]

Below Augusta and Hamburg, steamboats could navigate the Savannah River without river improvements. Although Georgia had jurisdiction over the Savannah River,[241] the second article of the 1787 convention between Georgia and South Carolina provided free and unimpeded access to South Carolina along the northern channel of the Savannah.[242] In practice this meant that South Carolina could initiate steamboat travel from Charleston to Hamburg, 140 miles above Savannah and directly opposite Augusta, Georgia. Such steamboat operations began in 1823.

The Steam-Boat *Hamburg*,
Captain HARVEY ; will be ready to receive Freight for HAMBURG, on *Thursday Morning*, at Fitzsimons's wharf. Those desirous of engaging Freight, can do so by entering it at the Scale House, Fitzsimon's wharf, where a Book will be left for the purpose.
march 15 3

Southern Patriot, March 27, 1823

Navigation on the Savannah River above Augusta and Hamburg would require more than clearing. As the Board of Public Works reported to the state legislature in 1820, "the difficult obstructions in the Savannah river can only be overcome by canals and locks."[243] Before this work could be undertaken, an understanding would have to be developed between the states of Georgia and South Carolina: "By the convention of 1787, the jurisdiction of the Savannah river is vested in the state of Georgia, and before we expend large sums in the improvement of its navigation, the subject of the toll to be levied upon the transportation of our produce, ought to be fully understood by both parties."[244]

As early as 1818 discussions were held between representatives of Georgia and South Carolina on improving the Savannah River for navigation, especially opening the Savannah River above Hamburg. In 1818 Wilson reported his efforts to the state legislature: "Agreeably to the resolution of the legislature in 1817, directing the Civil and Military Engineer of the state, to survey that part of the Savannah river contemplated in the report of the commissioners, and to report to the legislature at their next session, the facilities and obstacles to opening the navigation, and the probable expence of effecting it; I proceeded in June last, in company with the commissioners appointed by the state of Georgia, to examine that portion of the river."[245]

In 1818 Wilson and the Georgia commissioners began surveying the Savannah River above Hamburg. They had surveyed the river for a distance of only fourteen miles above Augusta and Hamburg when their work was interrupted by heavy rains.[246] Two years later, in 1820, Joel Poinsett, then president of the Board of Public Works, met with the Georgia commissioners on the subject of opening the upper Savannah River to navigation. No progress was forthcoming because "it appeared that they had not been vested with such powers as would enable the board to act in concert with them. The difficult obstructions in the Savannah river can only be overcome by canals and locks and the Commissioners on the part of Georgia are not authorized to purchase land, nor are they empowered to fix the rate of toll."[247]

Despite published reports that work would soon begin, nothing happened.[248] Probably the largest impediment to action was the view in South Carolina that improvement to navigation of the Savannah River would assist Charleston's chief competitor, Savannah.

Three years later, in 1823, Blanding reported to the state legislature that no progress had been made in developing an agreement with the state of Georgia over the Savannah River navigation:

> As yet, there has been no sufficient understanding between the two states on the subject, and the extent of the difficulties to be overcome on them is not well known. With the view of bringing the subject fully before the Legislatures of both states, I opened a correspondence through our executive with the governor of Georgia, and presented the following project of a convention between the two states on this important subject. I have the assurance of the Executive of Georgia, that it shall be presented to their Legislature at the next session. Should the plan meet the views of this state, its adoption here may be followed by the concurrence of Georgia, and the principles on which the works on this most important river are to be conducted, may thus be definitely settled.[249]

South Carolina and Georgia never reached an agreement on the navigation of the Savannah River and consequently no improvements to its navigation were initiated.

Building the State Road

Not all the internal improvements built by the state were waterways or canals. An important component of South Carolina's internal-improvement program was the State Road. The road had two principal parts: the 110-mile section between Columbia and Charleston and the 60 miles from the top of Saluda Mountain southeast to Columbia. Of the two sections, the Columbia to Charleston section was more difficult to build as it had to cross a number of major swamps requiring causeways and bridges.

The construction of the State Road, built between 1820 and 1829, has been greatly criticized. Historian Ulrich Phillips, for example, wrote: "The main state road was a demonstrated failure almost before it was completed. The road seems never at any time to have been in prime condition throughout its length; the tolls, amounting to nine dollars for a four-horse wagon on the round trip from Columbia to

Charleston and return, were more than the advantage was ordinarily esteemed to be worth."[250]

Some of this criticism was justified, but much of it was not. At the beginning of the state's internal-improvement program, there were few engineers available to supervise too many projects. Initially contractors on the State Road received very little supervision, and much of their work was unacceptable to the state. Building a road, particularly one crossing the swamps and rivers between Columbia and Charleston, was extraordinarily difficult. The working conditions were frequently adverse. Blanding provided an example:

> As soon as the president and engineer could examine the rout[e] between Charleston and Columbia, the work was begun at Goose Creek [above North Charleston] under Mr. Bee, at Wassamasaw Swamp [above Summerville] under Mr. Reardon, and at Dean Swamp [southeast of Holly Hill] under Col. Mellard. The work has been much interrupted by the rains; but such progress is made that it is believed three parts of the road will be finished before November [1821]. The swamp and bridge at Beaver Creek have been finished by Col. Myddleton. At Huckabuck Swamp, which is often covered five or six feet deep by the back waters from the Congaree, the high and extensive causeway was commenced by Mr. Anderson in April [1820]. The removal of his workforce at the approach of the sickly season, suspended his work for a short time, when it recommenced under Col. Myddleton with slaves, and is nearly completed and covered with gravel. A new Bridge on Town's place one hundred and four feet long, having but one arch, has been erected over the Congaree Creek [southwest of Cayce], and a similar bridge is preparing for Goose Creek, and will soon be up.[251]

Despite the initial difficulties in building this road from Columbia to Charleston, progress was made, and use of the completed sections of the road was heavy, as noted in this newspaper article from 1824:

> INTERNAL IMPROVEMENT. We learn that the contract for the State road, between Dean and Wassamasaw Swamps, (about 13 miles in distance) has been taken up, together with the other contracts, at prices so far below the estimates, that seven miles more of road, in addition to the Wassamasaw Causeway, have been let out. The construction of 20 miles of road is thus secured for the sum which it was estimated would be required for 13 miles of road. We also learn that the tolls at Dean Swamp for the two last months, considerably exceeded the interest of the work, and it has rented for the next year, for about 15 per cent. on the cost of the construction of the Bridge and Causeway.

Southern Patriot, February 3, 1824

An 1820 report on the Goose Creek portion of the new Columbia to Charleston Road (through Goose Creek Swamp, immediately above North Charleston) provides a description of what this road was like:

A gentleman who lately travelled over the Goose Creek Road, observed with great pleasure that about four miles of the turnpike, extending from the causeway towards this city, now nearly completed, has been as well executed, as the materials within reach would admit.... It is now laid off perfectly straight, about forty feet wide between the side ditches, with a sufficient elevation in the centre, raised with earth and brush-wood, to convey off all the rain water which falls, by a regular declivity into the drains. The trees have also been cut down for ten or twelve feet on the outer side of the ditches, in order to assist in keeping the highway dry, by the free access of sun and air; whilst at the same time it opens a magnificent avenue to the traveller, worthy the great emporium of whose commercial circulation it is destined to be a principal artery.[252]

The section of the State Road leading from Columbia to the Saluda Mountains, also called the Saluda Road, was less difficult to build:

The road through the Saluda mountains was examined early in June [1820], and its course determined on. The work was reserved, according to the direction of the legislature, as a retreat for the workmen on the rivers during the sickly season. An entirely new road was laid out and all the preparations made, so that the force commenced work of the 17th of July [1820]. Before the 1st of September [1820], more than half of this arduous undertaking was completed; it will be finished before the 15th of October [1820]. The road begins at the top of the Saluda mountain, and in ten miles passes down to the firm, level road, on the dry ridge between the Enoree and Tyger rivers, that extends in a direct course towards Columbia for fifty miles. No road is superior to it.[253]

The commissioners decided to relocate the workmen from the canals and waterways to the Saluda Road during the sickly season of 1820. About five hundred men began work on that road on July 17, 1820: "We have set every thing in operation on the mountain. After getting all our workmen housed, we commenced on the road on Monday, the 17th with about five hundred men; and they have already made a good impression on it. Those who have visited us have expressed their admiration at the extent of road already cut.... I have no doubt but that we shall complete the road before the fifteenth of October [1820]."[254]

Blanding provided the legislature with a description of the new road:

For [left blank in original] this road is cut in solid rock. It has [left blank in original] of stone wall to support it on the valley side, or to pass ravines. These walls average about 4 feet in thickness, and one of them is 23 feet high. There are 44 culverts or small bridges of stone flagged both at top and bottom. There are three arched stone bridges. At Gap Creek, the bridge is a Gothic arch, 15 feet high, and seven feet wide. The wall which connects the sides of the mountains, which inclose this stream, is 130 feet long, and from the water to the top of the parapet, has an elevation of 24 feet. The bridge over Saluda river has two elliptical arches, each 15 feet wide. The whole stone work is 60 feet long, and the elevation of the parapet wall above the water, is 15 feet. Over Hodge's creek, the bridge has one circular arch. The elevation of this bridge is 12½ feet, and the length of the bridge and abutments, 50 feet.

The road is usually 17 feet wide, but where on the steep sides of the mountain, much rock was encountered, it was cut 10 feet in the solid, and received as much additional width as the materials removed would form. Waggons may pass each other on all parts of the road at some point constantly in view. The great difficulty has been to reduce the road to a moderate angle. Many parts of the old road had an ascent of at least 12 degrees. There are but four places in the new road, where the angle is equal to 5 degrees, or a little less than one foot in ten. These four places in the whole, do not exceed 1¼ mile. About one mile more, has an angle of 4½ degrees, and no other part exceeds 3½ degrees. More than three miles of the road is almost a perfect level.[255]

Total costs for the development of the canals and river navigations discussed in this chapter were very large (see tables 4.3 and 4.4). The amount expended clearly exceeded the cost of the Santee Canal. It was also about the cost that Mills estimated for building a single canal from Columbia to Charleston. Within a short time

TABLE 4.3. Cost of constructing river-navigation canals, 1819–1828

Canal	*Dollars*
Drehr's Canal, Saluda River	78,139.63
Saluda Canal, Saluda River	161,415.46
Lockhart's Canal, Broad River	110,965.48
Lorick's Canal, Saluda River	2,168.84
Columbia Canal	206,268.69
Landsford Canal, Catawba River	129,843.61
Catawba Canal, Catawba River	164,992.26
Rocky Mount Canal, Catawba River	202,000.00*
Wateree Canal, Wateree River	159,854.77
Stono Canal at Ellicott's Cut	19,000.00*
Total	$1,234,464.86

*Estimated cost as of 1828

Source: *Report of the Superintendent of Public Works for 1828.*

TABLE 4.4. "General summary: total expenditures of the civil and military engineer, the Board of Public Works, and the superintendent of public works, 1819–1828"

Expenditure	*Under civil and military Engineer*	*Under Board Pub. Works*	*Under Sup Public Works*	*Balances due*	*Total expenditure on each class*
Class I. Roads		166,603.84	119,526.70	28,175.56	314,206.10
Class II. Canals	39,953.99	601,981.44	526,396.10	77,242.34	1,245,573.87
Class III. Rivers [below their falls]	58,351.94	106,761.13	47,623.66	5,051.50	217,788.23
Class IV. River [above their falls]			18,431.33		18,431.33
Salaries		29,571.27	31,693.60	1,544.78	62,809.65
Contingencies		6,166.87	21,617.70	2,060.71	29,845.28
Total under each direction	98,305.93	911,084.55	765,189.09	114,074.89	
Whole expenditure when balances are paid					$1,888,654.46

Source: *Report of the Superintendent of Public Works for 1828.*

frame, South Carolina had expended a huge amount of money on river navigation and on technologies then new to the United States. One has to be greatly impressed in the scope and vision that launched such an ambitious program.

Most of the money spent for internal improvements in South Carolina had been spent on canals and river-navigation schemes (see table 4.4). In return for this expenditure, South Carolinians expected a great deal of benefit.

5 ~ Operating the South Carolina Canal and River-Navigation System, 1819–1828

> That two thirds of all the market products of this State are raised within five miles of a navigable stream; and most of the other third within ten miles of a navigable stream; that fertility is only to be formed in the neighborhood of our water courses; that our streams run parallel to each other from the mountains to near the Sea-Board, as they approach it, all bend towards the same point; that traversing the country in a line with the Ocean, you meet navigable streams at every thirty miles.
>
> Colonel Abram Blanding at a town meeting in Camden, South Carolina; quoted in the *Southern Patriot,* December 5, 1820

SOUTH CAROLINA'S INTERNAL-IMPROVEMENT program had two objectives for the state's waterways: to clear the major rivers below their lowest falls for steamboat and canal-boat passage and to create a canal-boat navigation system above the falls with bypass canals above these and subsequent falls. Because it took time to build bypass canals around the obstacles above the falls, clearing the rivers below the falls was completed first. This allowed steamboats as well as other boats to navigate these rivers up to the fall line. In some cases steamboats began using South Carolina's rivers before the state engineer or Board of Public Works could clear them of logs and sandbars. Unlike the broad and wide Mississippi steamboats depicted in well-known engravings by Currier and Ives, South Carolina steamboats were long and narrow with a shallow draft. Their chief cargo was cotton, usually carried in towboats behind the steamboats. These early steamboats greatly reduced travel time from the upcountry to Charleston and proved very profitable to their backers. Unfortunately they had a tendency to run aground, sink, burn, or have their boilers explode. What we know of the early South Carolina steamboats comes from newspaper accounts of their unfortunate accidents, as well as advertisements for passengers and freight.

Steamboat Navigation on South Carolina's Improved Waterways, 1819–1828

Built in Charleston in 1819, the *Columbia* was one of the first steamboats used on South Carolina's waterways.[1] She had an overall weight of 129 $^{49}/_{95}$ tons; a three-foot six-inch draft; a deck eighty-seven feet long by eighteen feet wide, and a hold that was eight feet deep. She was powered by a single thirty-two horsepower steam engine.[2] The steamboat was constructed of live-oak and cedar frame, copper fastened.[3] In June 1819 the *Columbia* with her two towboats[4] began the first steamboat voyage from Charleston to Granby via the Santee and Congaree rivers. The Board of Public Works did not completely clear these rivers until 1821, and 1819 was a drought year, so the Santee River was extremely low, "as low as I ever knew it," said one observer. The *Columbia* ran aground a few miles below Lanud's Ferry on the Santee, but the crew was able to refloat her, and she proceeded on to Granby.[5] In 1822 the *Columbia* reappeared in the Charleston newspapers. She had been refurbished and was being used on the Charleston to Savannah run. An advertisement included information on its "accommodation of passengers,"[6] particularly ladies:

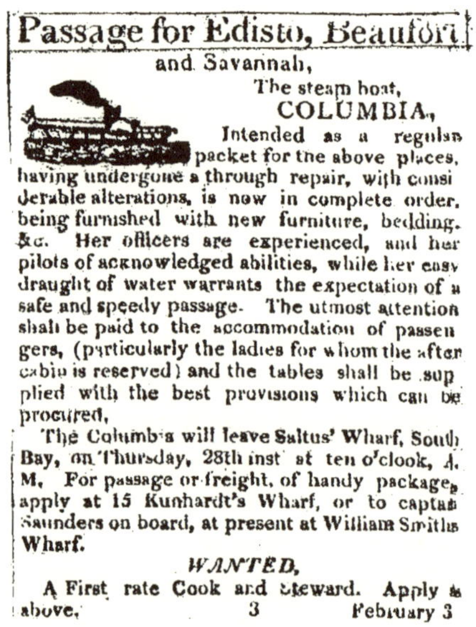

Passage for Edisto, Beaufort
and Savannah,
The steam boat,
COLUMBIA,
Intended as a regular packet for the above places, having undergone a through repair, with considerable alterations, is now in complete order, being furnished with new furniture, bedding, &c. Her officers are experienced, and her pilots of acknowledged abilities, while her easy draught of water warrants the expectation of a safe and speedy passage. The utmost attention shall be paid to the accommodation of passengers, (particularly the ladies for whom the after cabin is reserved) and the tables shall be supplied with the best provisions which can be procured,

The Columbia will leave Saltus' Wharf, South Bay, on Thursday, 28th inst at ten o'clook, A. M. For passage or freight, of handy package, apply at 15 Kunhardt's Wharf, or to captain Saunders on board, at present at William Smith's Wharf.

WANTED,

A First rate Cook and Steward. Apply as above, 3 February 3

Southern Patriot, February 23, 1822

In the following year, 1823, the *Columbia* was sold.[7] Then it again ran aground again and was again repaired. By June 1824 the *Columbia* was being used on the daily excursion runs from Charleston to Sullivan's Island.[8] In that same year the captain of the *Columbia,* Charles Clark, gave assurances to the public that the *Columbia* would not likely run aground again: "Her two recent detentions of one or two hours, occasioned by heating her condenser and getting aground, (occurrences common to steam boats) will not be likely again to recur: the recurrence of

Fig. 5.1. "Bird's Eye View of Charleston." From *Harper's New Monthly Magazine* (June 1857)

the former is effectually prevented, and the solicitude of the pilot, affords a very sure guarantee against the latter."[9]

Despite these assurances *Columbia* did run aground again later the same year, near Sullivan's Island on her voyage from Charleston to Georgetown and Cheraw. This accident proved disastrous. The *Columbia* began to take on water and subsequently sank, drowning four passengers.[10] She was later raised and refitted with a new bottom. By 1826 she was reported to be on the Cheraw-Georgetown-Charleston run via the newly cleared Great Pee Dee River.[11]

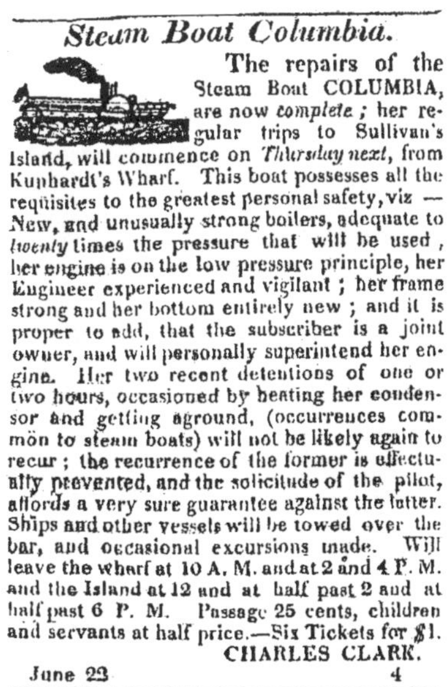

Steam Boat Columbia.

The repairs of the Steam Boat COLUMBIA, are now *complete*; her regular trips to Sullivan's Island, will commence on *Thursday next*, from Kunhardt's Wharf. This boat possesses all the requisites to the greatest personal safety, viz — New, and unusually strong boilers, adequate to *twenty* times the pressure that will be used, her engine is on the low pressure principle, her Engineer experienced and vigilant; her frame strong and her bottom entirely new; and it is proper to add, that the subscriber is a joint owner, and will personally superintend her engine. Her two recent detentions of one or two hours, occasioned by heating her condensor and getting aground, (occurrences common to steam boats) will not be likely again to recur; the recurrence of the former is effectually prevented, and the solicitude of the pilot, affords a very sure guarantee against the latter. Ships and other vessels will be towed over the bar, and occasional excursions made. Will leave the wharf at 10 A. M. and at 2 and 4 P. M. and the Island at 12 and at half past 2 and at half past 6 P. M. Passage 25 cents, children and servants at half price.—Six Tickets for $1.

CHARLES CLARK.

June 23 4

Southern Patriot, June 22, 1824

Another early South Carolina steamboat was the *South Carolina*. Built of live oak, cedar, and yellow pine, the *South Carolina* had a reported draft of three and a half feet[12]—ideal for South Carolina river navigation. Unfortunately on June 29, 1821, the *South Carolina*, while moored at Magwood's wharf in Charleston, "was almost completely enveloped in flames" and "burnt nearly to the water's edge, she was scuttled and almost immediately sunk."[13]

The ruins of the *South Carolina* were sold, and the steamboat was rebuilt.[14] In 1822 the reconstructed *South Carolina* made the voyage from Granby down the Congaree and Santee rivers to Charleston in only four days. Enthusiastically reporting on the voyage, a Charleston newspaper stated: "The steam boat *South Carolina*, arrived this morning from Granby, in the remarkably short passage of four days, and before the manifest of her cargo has been received in this City, by the mail. We thus see the facility of this system of navigation, and its benefit to our City realized on every side. We hope the patronage it is to enjoy will be large and liberal."[15]

Today four days to travel from Granby, just south of Columbia, to Charleston, a distance of approximately 110 miles, does not seem like a "remarkably short passage." However, it was in the 1820s. It represented one half the time that canal boats then took to cover the same distance and was also faster than the mail coach from Columbia. The return voyage upriver to Granby by steamboat was much longer, requiring twelve days. Total time for a round trip took sixteen days or less,[16] but this was also shorter than the voyage aboard a canal boat, which took thirty days for the round trip from Granby to Charleston.[17]

Freight for Granby.

Freight will be taken for Granby, and along the river, on board two CANAL BOATS, lying at Fitzsimon's wharf, if applied for immediately.

JOHN SCHULTS.

December 24 2

Southern Patriot, December 24, 1822

For Columbia.

The Steam Boat SOUTH-CAROLINA, with her Tow Boat, Wm S. Huntermaster; is now ready to take in, and will leave here on the 11th inst. For Freight, apply to

MICHAEL LAZARUS.

December 9 2

Southern Patriot, December 9, 1825

The rebuilt *South Carolina* was sold in 1823 by the Charleston and Columbia Steam Boat Company.[18] For several years she continued hauling cotton from Granby and Columbia to Charleston and merchandise back via the cleared Congaree and Santee rivers. Two of her towboats could haul 750 to 800 bales of cotton each while the other two could carry 350 to 400 bales each.[19] Early in 1826 the *South Carolina* was reported on the Hamburg-Beaufort-Savannah-Charleston run.[20] That year the Savannah River was extremely low, and it is likely that the *South Carolina* was damaged there, for in October of 1826 she was reported as having undergone complete repairs and was then reported on the Cheraw-Georgetown-Charleston run.[21]

The steamboat *Maid of Orleans,* was operating in 1822 between Charleston and Cheraw on the Great Pee Dee River.[22] She had been constructed in 1819 "of the best materials, copper boilers, and from her light draft of water well calculated as a freight boat for river navigation."[23] In 1823 she was offered for sale, and by 1824 she was being used in the Charleston to Hamburg and Augusta run via Beaufort and Savannah.[24]

Extract of a letter from Mr Charles Clarke, Engineer of the steam boat Hamburg, to a gentleman in Augusta, dated on board, May 28th, 1825.

"You have heard before this of our sinking the steam boat. We have damaged the most valuable cargo that we have had on board this spring, consisting of sugar, coffee, dry goods, hats and shoes.—By the assistance of a large box pump, which I made of planks, wrought by twelve men, in addition to five other pumps and one barrel, and all worked with ut a moment's intermission for about 30 hours, we were enabled to free her and stop the leak. I have now got my pump connected to the Engine in case of another accident; I calculate it will deliver 60 cubic feet of water per minute."

Southern Patriot, June 5, 1826

The steamboat *Charleston,* built in 1817, was 160 tons with a coppered hull over oak and cedar and equipped with heavy copper boilers.[25] In 1820 she was reported hauling freight to Columbia and Granby[26] and sailing to Beaufort and Savannah.[27] By 1821 the *Charleston* had joined the *Columbia,* under Captain J. Bonnell, in operating between Charleston, Beaufort, and Savannah.[28] In 1823 *Charleston* appeared near Granby,[29] traveling on the Congaree and Santee rivers with the steamboat *Patent* and others.[30] Later in 1823 the *Charleston* was operating from Charleston to Cheraw when she ran one of her towboats upon a snag in the Great Pee Dee River, immediately sinking the towboat and losing the cargo.[31] By April 1826 her two towboats were sold.[32] Later that same year, the *Charleston* was auctioned off.[33] It is likely that she was then extensively refitted, for in September 1826 the Charleston papers reported the launching of the steamboat *Charleston.*[34]

Even without clearing, the Savannah River was navigable as far upriver as Hamburg, across from Augusta, Georgia. By 1823 the steamboat *Hamburgh* was operating between Charleston and Hamburg via Beaufort and Savannah under the command of Captain Harvey.[35] In June 1826 she sank in the Savannah River, near Augusta and Hamburg,[36] but by October of that year she had been raised and refitted. She was reported as having undergone a complete repair, "and is now as safe a boat as any in the trade."[37]

For Augusta and Hamburg,
Via Beaufort and Savannah,

The Steam Boat HAMBURGH, Capt. Andrew Boyden, will commence loading for the above places *Tomorrow Morning* at *Magwood's wharf*, and will be despatched full or not full, in a few days

The Hamburg has undergone a complete repair, both in her hull and machinery, and is now as safe a boat as any in the trade—has *copper boilers*, and the best engine in the Southern States.

Safe and convenient *Brick Stores* are obtained on *Magwood's wharf*, for the reception of Goods, free of storage, intended to be shipped by her, and will be shipped free of any expense except wharfage and drayage.

The Cabin of the Hamburg is neatly fitted up, and passengers will find good fare, and every attention paid them by Capt. Boyden.

Goods for the interior of Georgia or South Carolina, shipped by her, will be received by the agents of the Boat, Messrs. Pilot & Le Barbier, in Augusta, and forwarded to the owners, free of any charge except those actually incurred. No Powder will be permitted to be shipped on board the boat, but a *new Powder Boat* will be ready the next trip of the Hamburgh. JOHN ROBINSON, SON & CO

Oct 2 *Agents Steam Boat Hamburg.*

Southern Patriot, October 13, 1826

Later in 1824 the steamboats *Wm. Lownes* and *Commerce* also began operating between Charleston and Hamburg using the Savannah River.[38] The *Commerce* was sold in March 1826 along with "one half the Steam Boat *Augusta.*"[39] It is not clear what happened to the other half of the *Augusta.* In 1824 these steamboats were joined by the *Henry Schultz,* recently built in Philadelphia by the firm of Rush & Mullenburg. The *Henry Schultz* was described as having, "accommodations [that] are very superior, and her Machinery of the most approved kind."[40] This steamboat was powered by a sixty-horsepower steam engine (large for vessels of the time) fired from four iron boilers measuring twenty-three feet in length and thirty inches in diameter.[41] Unfortunately the *Henry Schultz* only survived a short time, wrecking early in 1825 on the Savannah River near Augusta. The underwriters sold the remains, including the steam engine, at an auction

Valuable Steam Engine at
AUCTION.

On MONDAY, the 9th May next, at 12 o'clock, will be sold at the wharf, in Augusta, on account of the underwriters and all concerned,

The Wreck of the steam boat HENRY SHULTZ,

As she now lies, together with her Rigging, Sails, Anchors, &c. ; and at the same time and place the Engine of said boat, made by Rush & Mullenburg, of Philadelphia, upon the high pressure principle, full 60 horse power, with four iron Boilers, measuring 23 feet in length and 30 inches diameter, and Furnace complete, not one year old, and considered by the Engineer, to be uninjured by the fire or explosion.

The Engine is thought to be complete, with the exception of one of the main Shafts being injured. Two others however, are in Charleston, made on an improved principle to those now on board, and will be sold as part of the machinery.

Conditions cash. Darien and North Carolina money taken at par in payment.

☞ The Charleston Courier and Patriot, and Savannah Georgian and Republican, are requested to copy this advertisement 5 times, and forward their accounts to the agent of the boat, in Augusta.

April 27 6 [A. 30]

Southern Patriot, May 4, 1825

MARINE NEWS.
PORT OF CHARLESTON.
HIGH WATER THIS DAY : : : . : : : . : -7H 49M

ARRIVED THIS DAY.

Steam boat Edgefield, Sassard, Augusta and Hamburg, 6 days. 900 bales Cotton, and 30 hhds Tobacco. To R M Hamilton & Co R Martin & Co J Magrath, A D Torre, L C Cantelou, Montgomery & Platt, Boyce & Henry, Parish, Corning & Co J Robinson, Son & Co B Saltar, T W Bacot, Hyde & Cleveland, Tupper & Kimball and S B Jones. *Passengers,* Mr and Mrs Hamilton & child, Miss Bythwood, Capt Lubbock, Messrs Smith, Darby and Goff. Passed the steam boat Hamburg, at Savannah on the 3d inst ; and steam boat Augusta, Green at Beaufort, where she will be detained a few days to repair some damage sustained by her engine.

Southern Patriot, February 7, 1826

held on May 5, 1825.[42] Other steamboats on the Charleston to Hamburg run included the *Pendleton,* the *Edgefield,*[43] the *General Greene,* and the *Macon.*[44]

South Carolina steamboats were owned and operated by companies established for that purpose. One of the earliest was the Congaree and Santee Boat Company, established in Charleston in 1819.[45] They owned the *South Carolina* until after it burned in June 1821, selling it on August 13 along with the company's four towboats and two slaves who acted as boat hands on the steamboat.[46] This company was superseded by the Charleston and Columbia Steam Boat Company, organized to operate steamboats on the Granby to Charleston run.[47] By 1825 the steamboat *Spartan* was also reported to be operating between Granby and Charleston.[48]

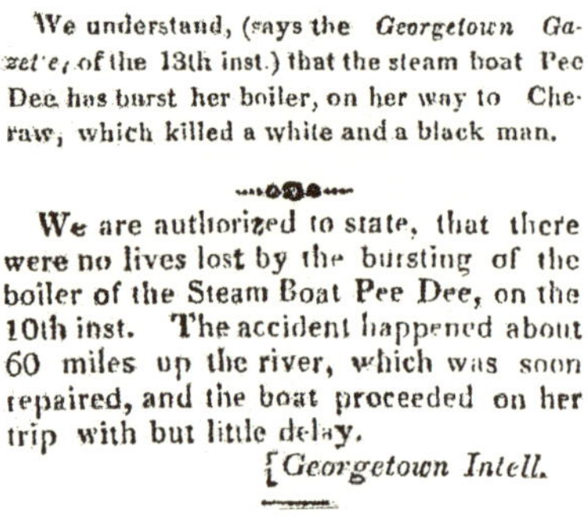
We understand, (says the *Georgetown Gazette*, of the 13th inst.) that the steam boat Pee Dee has burst her boiler, on her way to Cheraw, which killed a white and a black man.

We are authorized to state, that there were no lives lost by the bursting of the boiler of the Steam Boat Pee Dee, on the 10th inst. The accident happened about 60 miles up the river, which was soon repaired, and the boat proceeded on her trip with but little delay.
[*Georgetown Intell.*

Southern Patriot, January 16 and 26, 1826

Another early South Carolina steamboat company was the Pee Dee Steam Boat Company, whose boats operated on the Cheraw-Georgetown-Charleston route. One of the steamboats it used on the Great Pee Dee River was called the *Pee Dee.* In January 1826 the Charleston newspapers reported that the *Pee Dee* had burst its boiler and that two men were killed.[49] Later it was reported that no lives were lost.[50] Two months later the *Pee Dee* was auctioned off.[51] It was then used on the Hamburg-Beaufort-Savannah run.[52]

Despite the bursting of the *Pee Dee*'s boiler, the Cheraw-Georgetown-Charleston run was profitable, and the Pee Dee Steam Boat Company declared a 12 percent dividend as early as 1821.[53] By October 1826 the *Cheraw Spectator* was reporting that the steamboats *General Greene, South Carolina,* and *North Carolina* were going to be used on the Cheraw-Georgetown-Charleston run during the coming winter of 1826–27.[54]

In 1819 the trip on the Great Pee Dee River from Cheraw to Georgetown, a river distance of some two hundred miles, was expected to take a steamboat no more than seven days, substantially less than the twenty to twenty-five days required by other boats then operating on the Great Pee Dee.[55] The following year the Charleston newspaper reported that a steamboat took only two days to travel from Chatham (Cheraw) down the Great Pee Dee River, past Georgetown, and down the coast to Charleston.[56]

By 1823 the citizens of Camden were proposing to establish the Camden Steam Boat Company. Its boats would operate on the newly cleared Wateree and Santee rivers between Camden and Charleston.[57]

By 1828 newer steamboats had been added to South Carolina's waterways. The *Governor Taylor* was described as "elegant," "comfortably fitted up for passengers," and in "every way calculated for speed." She was built "solely for the Columbia trade, and is expected to make two trips per month."[58] Also in 1828 the *General Jackson* was launched from the Charleston shipyard of James Marsh and Son. The *General Jackson* was intended for the Augusta and Hamburg trade and could carry nine hundred to one thousand bales of cotton.[59] In the same year the older steamboats *Augusta* and *Commerce*, then partially dismantled, were put up for sale.[60]

Steamboat navigation was not only proving to be fast but also to cut the cost of freight shipment by one half or more.[61] In 1822 the Board of Public Works reported:

To realize the extent of benefit resulting from the general introduction of Steam Boats suited on the rivers to them and from the substitution of small boats on our upper rivers in room of wagons, it is only necessary to go into the following calculations....

Now, where these boats have been introduced, it has lowered the freight before paid to pole boats about 100 cents on a bale of cotton. But the whole 130,000 bales [of cotton] that may reach the Charleston market, do not descend the whole distance, and therefore, the average saving on a bale may not exceed 50 cents, or $65,000 in the whole: an equal saving may attend the up freights. The result of all this would be.—

Saving on the upper river, ... $105,000
Do. by steam boats on the lower rivers, ... 130,000
Whole saving per annum .... $235,000[62]

Besides river navigation, steamboats were also used along the coast, especially on the run from Charleston to Beaufort, Edisto, and Savannah. Passage on these steamboats was not inexpensive; a one-way trip for a person traveling from Charleston to Beaufort cost nine dollars, and a one-way trip from Charleston to Savannah cost thirteen dollars—at a time when skilled workers were receiving approximately a dollar a day in wages.[63]

Besides improvements to rivers that served the interior of the state, Blanding, as superintendent of public works, undertook improvements on coastal routes used by steamboats and other vessels: "The Islands which stretch in continued succession from Charleston to Savannah, afford an inland navigation between the two cities, which is capable of being rendered perfectly good for steam-boats. There are, however, several places where there are obstructions, which must be removed, before the passage from the Savannah river can be entirely inland. The principal obstruction is at Wall's cut where steam boats cannot now pass; and considerable labor

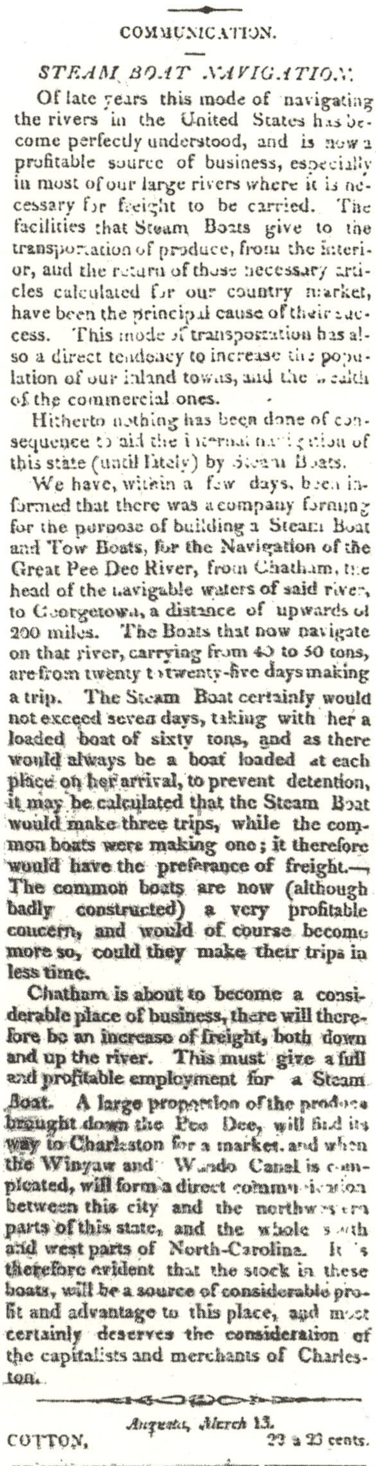
COMMUNICATION.

STEAM BOAT NAVIGATION.

Of late years this mode of navigating the rivers in the United States has become perfectly understood, and is now a profitable source of business, especially in most of our large rivers where it is necessary for freight to be carried. The facilities that Steam Boats give to the transportation of produce, from the interior, and the return of those necessary articles calculated for our country market, have been the principal cause of their success. This mode of transportation has also a direct tendency to increase the population of our inland towns, and the wealth of the commercial ones.

Hitherto nothing has been done of consequence to aid the internal navigation of this state (until lately) by Steam Boats.

We have, within a few days, been informed that there was a company forming for the purpose of building a Steam Boat and Tow Boats, for the Navigation of the Great Pee Dee River, from Chatham, the head of the navigable waters of said river, to Georgetown, a distance of upwards of 200 miles. The Boats that now navigate on that river, carrying from 40 to 50 tons, are from twenty to twenty-five days making a trip. The Steam Boat certainly would not exceed seven days, taking with her a loaded boat of sixty tons, and as there would always be a boat loaded at each place on her arrival, to prevent detention, it may be calculated that the Steam Boat would make three trips, while the common boats were making one; it therefore would have the preferance of freight.—The common boats are now (although badly constructed) a very profitable concern, and would of course become more so, could they make their trips in less time.

Chatham is about to become a considerable place of business, there will therefore be an increase of freight, both down and up the river. This must give a full and profitable employment for a Steam Boat. A large proportion of the produce brought down the Pee Dee, will find its way to Charleston for a market, and when the Winyaw and Wando Canal is completed, will form a direct communication between this city and the northwestern parts of this state, and the whole south and west parts of North-Carolina. It is therefore evident that the stock in these boats, will be a source of considerable profit and advantage to this place, and most certainly deserves the consideration of the capitalists and merchants of Charleston.

Augusta, March 13.

COTTON, 23 a 23 cents.

Southern Patriot, March 9, 1822

is required to open a passage from the Savannah into New River. This work is of vast importance, for without it, boats descending the Savannah, and proceeding to Charleston, are obliged to go out to sea, and are much exposed."[64]

Steamboats could not be used above the falls on rivers. They were too large to use in the newly constructed bypass canals and in the Santee Canal. Despite their minimal draft of three and a half or four feet, they were frequently not able to navigate the seasonal low water of South Carolina's rivers during the summer and autumn or during periods of drought.

Periods of low water plagued navigation on most South Carolina rivers. For example in 1828 the Congaree River was entirely closed to navigation because of low water.[65] Newspaper accounts reported similar situations for the other rivers in the state.

Although data is lacking for how many months in a year that steamboats could be operated on South Carolina rivers, a newspaper article about steamboats at that time on the Alabama River indicated that low water on the river limited steamboat travel for half the year.[66] South Carolina steamboat operations probably had a similar record.

The Augusta Chronicle received last evening, says, "The River is now nearly as low as it has been at any time during the late summer The Steamboat Hamburg is aground below the Bridge, and the Macon is in a similar predicament at Bagg's bar."

The Steam boat North Carolina has reached within six miles of town, and as the river is now rising she will undoubtedly get up this day.

The fine new boat building by Messrs. J. & J. H. Townes of this place, will be launched on Tuesday next. This boat is intended for the navigation of the Pee Dee, is 104 feet keel, and is we understand the largest boat ever built on this river.

Cheraw Spec. Nov. 30.

Southern Patriot, November 4 and December 3, 1827

Team Boat *Yadkin.*

This beautiful boat, lately built at Cheraw, of easy draught of water, so as to admit of her running at all seasons, has now commenced running from Cheraw to Georgetown. The subscribers propose to their friends in North and South Carolina to freight their goods and produce at the usual steam boat prices, with the exception of salt. The boat will be under the particular care and direction of either of the subscribers, Henry H. Miller, at Cheraw, Benj. King, Georgetown, and B. D. Lazarus, and Jno. Magrath at Charleston.

J. LAZARUS & CO.
T. ADDISON.

May 14 3

Southern Patriot, May 16, 1823

Fig. 5.2. Although it dates from well after the canal period in South Carolina, this rare illustration of a Charleston river craft, from an 1876 issue of *Frank Leslie's Popular Monthly,* depicts a boat that was typical of those used on the state's canals throughout the canal boom.

Besides steamboats, other vessels traveled the recently cleared rivers. Boats powered by horses and mules were used between Cheraw and Georgetown on the Great Pee Dee River and probably elsewhere. Traditional South Carolina boats—locally known as Camden, Congaree, or Wateree boats—continued to be used in rivers and canals after the rivers were cleared for deeper-draft vessels. Typically these boats were sixty to seventy feet long and fifteen to twenty feet wide; they carried fifteen to twenty-five cords of wood or the equivalent.[67] These boats were probably local forms of the pettiauger (also spelled *periaugua, piragua,* or *pettyagra*), a boat that was widely used throughout eighteenth- and nineteenth-century America. These unpowered, flat-bottomed boats were typically poled and sometimes equipped with sails.

Another vessel used on South Carolina waterways was an ancient type of North American river craft, the bateau. Typically ten to sixteen feet long, a bateau would carry sail, oars, and an awning to protect its cargo.[68] Canoes were also used on South Carolina's waterways. Much larger than contemporary canoes, these early-nineteenth-century canoes were thirty or more feet long and five or more feet wide with space for five or six oarsmen, usually slaves.[69]

Canal boats used on South Carolina's rivers and canals usually carried 100 to 120 bales of cotton downriver and assorted manufactured goods upriver.[70] Manned by slaves, they were usually poled and towed by their crews through South Carolina canals, such as the Santee Canal.

New Pettiauger to be ſold cheap.

TO be ſold, at private ſale, a New CAMDEN BOAT, built of the beſt Materials, her timbers of Mulberry and Saſſafras, well bolted and ſpiked, will carry about 1700 buſhels of Corn or 25 Cords of wood. For terms apply to

BRIAN CAPE & SON.

January 20.

FOR SALE,

An Excellent

Congaree-Built BOAT,

Burthen 17 Cords of Wood, of an eaſy draught of water, well calculated for the river trade, with Anchors, Cables, Sails, &c. As the proprietors have no further occaſion for her, ſhe will be ſold cheap, and payments to ſuit the purchaſer. Apply to

Shackelford & Ham.

No. 10, Gaillard's Wharf.

City Gazette, January 21, 1797, and March 6, 1800

A CANOE.

TAKEN up, a drift, near Fort-Johnſton, a Canoe, thirty-three feet in keel, and five feet five inches wide, rows with ſix oars, and appears to have had her bottom pitched. She has no other particular mark. Whoever owns her may have her again, by paying for advertiſing her, and the charges according to law, allowed for trouble. Apply at Fort-Johnſton, to

John Bevan.

March 7. 3 eod

City Gazette, March 7, 1800

A Canal Boat adrift.

Drifted from one of tle wharves on Saturday last, a CANAL BOAT, sides painted red, and hatches covered, with tarpouling nailed on them A suitable rewad will be given to any person who may pick he up, and give information at No 6, Crafts' Soth Wharf.

February 27

Southern Patriot, March 3, 1823

Travel times for canal boats descending the rivers and canals of South Carolina were long, typically from nine to eleven days from Camden or Columbia to Charleston on the Wateree and Santee rivers and down through the Santee Canal. The time from Stateburg, also via the Wateree and Santee rivers, was a little less. Canal boats ascending South Carolina's rivers and canals had to be poled against the current, so upriver travel times were usually twice as long as downriver.

Canal boat crews were large, usually from six to seven slaves. Owners allowed these crews a great deal of independence because they were away from their plantations for weeks at a time. These slaves were highly valued for their special knowledge of river conditions and navigation. Typically boat crews were sold or rented as a team.

By 1822 the Board of Public Works was reporting that more than 1,280 miles of rivers and waterways in South Carolina were open to steamboat navigation and

TABLE 5.1. Mileage of navigable South Carolina waterways reported in 1822

	By land	*By water*
The extent of Steam Boat Navigation within the state, on rivers and inland passages, on which the state has made expenditures, measuring the distance by land	450 miles	728 miles
The extent . . . on rivers and inland passages, on which there has been no expenditure by the state	420 miles	552 miles
Whole extent by steam boat navigation	870 miles	1,280 miles
Extent of small boat navigation, within the state, on which there have been expenditures,	398 miles	580 miles
Extent of small boat navigation within the state, on which there have been no expenditures,	451 miles	622 miles
Whole extent of inland navigation within the state	1,719 miles	2,418 miles
Whole extent of inland navigation in North-Carolina, connected with our rivers,	340 miles	
Whole extent by land,	2,059 miles	

Source: *Report of the Board of Public Works for 1822.*

another 1,200 miles were open for small-boat navigation (see table 5.1).[71]

For Sale,
A new CANAL BOAT, built of the best materials, carrying from 9 to 10 cords of wood. For terms, apply to
David Adams & Son.
march 6 stu *Doyle's Wharf.*

Southern Patriot, March 10, 1818

Singleton's boat, 8 days from Stateburg. 94 bales cotton, and 150 bushels ground nuts—to J H Merritt, D Goodman, and F W Sollee.
Cyple's boat, 11 days from Camden. 101 bales cotton—to W Clarkson.
Purvis' boat, 11 days from Columbia. 93 bales cotton—to W Clarkson.
Robinson's boat, 9 days from Camden. 97 bales cotton—to J Robinson,
Berry's boat, 10 days from Camden. 105 bales cotton—to J Robinson,

Southern Patriot, November 25, 1822

CANAL BOATS TO CHARLESTON

The Saluda, Broad, Congaree, Wateree and Catawba river navigations were improved to funnel agricultural goods, particularly cotton, to Charleston via the Santee Canal. Unfortunately, as the ambitious South Carolina internal improvement began, the state was hit by droughts in 1816 and again in 1818–19. Critical sections of the Santee Canal dried up. Level 2–3, between locks 2 and 3 in the northern section, which relied on water from Greenland Swamp (through which the surveyor Mouzon had originally recommended the Santee Canal be constructed as his first route for the canal), was completely dry. Level 3–4, the summit level, was also dry. All the canal's four supplying reservoirs failed. Level 4–5 was dry as well, its reservoir having also failed.

John Christian Senf, chief engineer for the Santee Canal, has received much of the blame for the Santee Canal remaining impassable to traffic during the droughts of 1816 and 1818–19, particularly from historian Frederick A. Porcher. Porcher suggested that, if Senf had selected one of Mouzon's routes for the Santee Canal, he would have avoided the impact of the droughts.[72] It was true that the route that Senf selected, beginning at White Oak Landing two miles east of Greenland

Swamp, was five feet higher than Mouzon's route through the Greenland Swamp, but during these drought years Greenland Swamp was also dry. No route selected between the Santee and Cooper rivers would have avoided the impact of these two drought periods. Only a much deeper canal or a canal mechanically supplied with water would have remained in operation.

In order to counteract the effects of the drought, the company purchased two steam-powered pumps to provide water to the dry levels of the canal. One was placed at lock 2 and the other at lock 3. Neither worked well. According to the 1881 army report on their operation on the Santee Canal: "The first [pump], after working for some time, barely supplied the water lost by leakage, and the second did scarcely as well. They were both soon given up."[73]

During those drought years, cotton still needed to be delivered to Charleston for export. Planters found alternative means. Porcher wrote, "I remember when a boy . . . all the roads to Charleston whitened by the tents of the wagons as they slowly pursued their toiling way to and from the metropolis; and then did the unfortunate spirit of public improvement get possession of our Legislature and make turnpikes for these cotton laden wagons."[74]

The Santee Canal was the weakest link in the state's expensive and extensive state's river-navigation system. Because its closing during droughts threatened much of the system, the Board of Public Works and its successor, Superintendent of Public Works Abram Blanding, sought to improve the Santee Canal. In May 1823 Blanding[75] inspected the Santee Canal to develop recommendations for improving it. He was accompanied by Santee Canal superintendent Henry Glindkamp[76] and one of the company's directors, Henry Ravenel.[77]

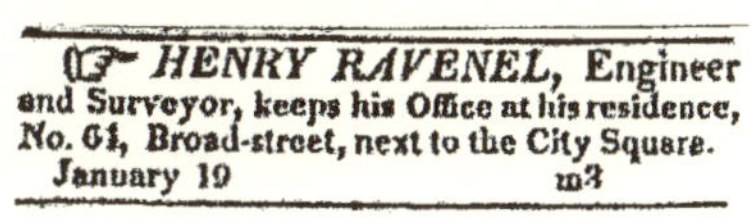
☞ HENRY RAVENEL, Engineer and Surveyor, keeps his Office at his residence, No. 61, Broad-street, next to the City Square.
January 19 m3

Southern Patriot, January 19, 1824

Blanding found two major defects with the Santee Canal: that it dried up during extreme droughts and that the locks were too small to be used by the larger riverboats. He rejected the idea of rerouting the canal through Greenland Swamp, as originally proposed by Mouzon:

> Opinions have been expressed by some, that the summit level should be changed and carried through Greenland Swamp. What is to be gained by this? The only difference in level between the ridge on that rout[e], and the ridge where the canal now exists, is five feet. The whole length of the new canal will be about 8 miles; and at last, the prospect of water from the surface of the earth is not rendered any better; for this swamp in 1816, 1818, and 1819, was entirely dry. And of course the only water to be depended on, for the new canal, must be derived from percolation through its banks and bottom. How deep the excavation must be to obtain this supply, is perfectly uncertain; probably, however, as deep as the present summit canal. To change the canal to Greenland swamp, will, therefore be attended with immense expense, without any certainty of attaining the desired object.[78]

In order to solve the first defect, too little water, Blanding proposed that the deepest areas of the canal—some fourteen and in some cases seventeen feet below grade—be cut deeper by an additional four feet. He estimated that fifty thousand dollars would be needed to do this and that, once dug, the canal would have sufficient water on all levels, even during drought years.[79] Blanding recommended:

— lowering the summit level, level 3–4, between locks 3 and 4 (Frierson's Lock), by four feet. Blanding noted that even during the droughts of 1816 and 1818–19, the summit canal had eighteen inches of water in its bottom, percolating in from the surrounding soil. Blanding calculated that, by deepening the prism, the water table of the entire country would provide sufficient water to the summit level of the canal.
— lowering level 4–5, between locks 4 and 5, by four feet eight inches to allow the water from Frierson's and Gaillard's springs, now above the canal, to flow into level 4–5.
— lowering level 6–7, between locks 6 and 7 (Black Oak Lock), by two feet four inches so as to allow water from Ravenel's spring to flow into it.[80]

In order to solve the second defect, that the locks were too small, Blanding proposed enlarging the locks to permit boats with sixty-five-to-seventy-foot keels and sixteen-foot beams. (At that time the Santee Canal could pass boats that had fifty-six-foot keels and nine-and-a-half-foot beams.) This enlargement would permit boats carrying up to 250 bales of cotton to pass through the canal (well more than the existing limitation of approximately 120 bales).

Blanding also proposed a number of other plans for improving the Santee Canal, such as installing various devices, including a steam-powered pump and water-saving locks. None of these plans were ever funded by the South Carolina legislature. This was partly because the Santee Canal was owned by a private company, partly because rain ended the drought of 1818–19 and resurrected canal traffic, and partly because there was a growing disillusionment with river navigation in general throughout South Carolina.

Once the rains resumed, traffic on the Santee Canal increased to 750 boats per year carrying 70,000 bales of cotton in 1830.[81] Blanding had estimated that the total of cotton delivered to Charleston in 1823 was 140,000 bales.[82] The amount of cotton shipped through the Santee Canal for that year is not available but it appears that the Santee Canal was shipping a significant portion of the entire cotton crop being delivered to Charleston. These were the boom years for the Santee Canal.

The high point for the Santee Canal was 1830. Commerce declined thereafter. The rivers above the Santee Canal began to be blocked again by sandbars and logs, limiting the traffic that fed downriver traffic to the Santee Canal. Within a decade after construction, the canals on the rivers above the Santee Canal were abandoned because of a lack of traffic. Roads, particularly the new State Road between Columbia and Charleston, took away much of the traffic. And finally the railroad came to South Carolina.

In the 1830s the state-constructed bypass canals began to be abandoned. Of the river-navigation network that fed into the Santee Canal, the Saluda bypass canals had been the first to be constructed, but the Saluda navigation system and its canals had never carried much traffic. During the late 1820s and early 1830s, the small amount of traffic there declined sharply. Drehr's Canal transported 708 bales of cotton in 1826,[83] 578 in 1827,[84] and 41 in 1828.[85] The canal's tolls came to $92.21 in 1827, and $7.54 in 1828; the canal apparently closed from 1831 to 1833 and received $6.77 in tolls in 1834.[86] The Saluda Canal shipped 712 bales in 1826,[87] 590 in 1827,[88] and 108 in 1828. Its tolls came to $125.17 in 1826 and $113.75 in 1832; by 1834 it was yielding no revenue through collected tolls.[89] It appears that these bypass canals and the navigation channels on the Saluda River were completely abandoned by 1837[90] (see table 5.2).

The Catawba River navigation was the last to be completely opened, owing to the difficulty in constructing the Rocky Mount Canal, which was not completed until 1828.[91] The Catawba River navigation experienced little traffic, and that traffic declined during the early 1830s. Anticipated freight from North Carolina never materialized. There is no record of tolls having been received at the most northern of the Catawba bypass canals, the Landsford Canal. The southern bypass canals were also little used. Tolls for the Catawba, or Fishing Creek Canal, were $43.99 in 1831, $51.10 in 1833, and $16.71 in 1834. The last year of operation was probably 1837.[92] Rocky Mount Canal tolls peaked at $312.69 in 1831 but fell to $65.12 by 1833. There is no record of tolls being collected after that year.[93]

The Wateree Canal, on the Wateree River below the Catawba, was more successful. In 1827, 3,144 bales of cotton were shipped through this canal.[94] But then traffic declined. Tolls dropped from $276.41 in 1831 to $154.23 in 1834.[95] Records

TABLE 5.2. Annual cotton traffic and tolls on select canals

Canal	*Year*	*Boats Carrying Cotton*	*Bales of Cotton*	*Tolls Received*
Santee Canal	1830	750	70,000	n/a
Columbia Canal	1827	473	45,612	$2,993.08
Saluda Canal	1827	23	590	$125.17
Drehr's Canal	1827	21	578	$32.21
Broad River Canal	1827	89	2,544	$331.40
Wateree Canal	1827	68	3,144	$327.84

The Broad River Canal was also called Lockhart's Canal.

Source: *Report of the Superintendent of Public Works for 1827.*

indicate that expenditures for the canal for the period 1831 to 1834 exceeded tolls received by $3,210.19.[96]

For a time the Wateree Canal, plus the extension of the State Road to Camden, brought prosperity to that area. An 1828 newspaper article reported:

> CAMDEN, DEC. 6.
>
> *Business of our Town.*—We apprehend that few towns of its size in the country, exhibit a busier aspect than Camden at the present time. Broad street has for several days been literally *choked* with wagons, carts, and all the various vehicles, ever made use of, in the conveyance of produce and merchandize. On Wednesday, eighty four wagons, loaded with Cotton, Corn, Wheat, Flour, Iron, &c. were counted in the street at one time, and upwards of one hundred and fifty arrived in the course of that day. Thirty-three boats arrived on Tuesday, and nearly that number are at this moment loading with Cotton for Charleston: in short all is activity, bustle and business, and our river craft only lack masts, to give our landing the appearance of a lively seaport town.

Southern Patriot, December 9, 1828

But by 1838 the Catawba/Wateree navigation system had been totally abandoned, superseded by roads and the railroad.[97]

The most successful of the bypass canals built by the state was the Columbia Canal. Bales of cotton shipped through this canal grew from 21,863 in 1825,[98] to 45,612 in 1827,[99] to 48,874 in 1829, and to a peak of 66,597 in 1833.[100] A substantial portion of the cotton arriving at Charleston was shipped through the Columbia Canal.[101] Revenue received was substantially ahead of costs.

> Abstract from the books of the operations of Columbia Canal, from 1st October, 1826, to 1st July, 1827.

	Cotton Boats	other Boats	whole number of Boats	Bales of Cotton.	Tolls.
1st quarter	126	99	225	11 915	$ 704 46
2nd do	199	216	415	20 485	1326 09
3rd do	140	147	287	13 055	902 33
Total	465	462	927	45 455	$2932 88

> It will be seen by the above statement, that 45,455 bales of cotton passed the Columbia canal in the last nine months. During the same period, the shipments by the Columbia merchants, from Granby below the canal, has been from 9 to 10 thousand bales; so that the whole country above Columbia and connected with it in trade, has sent this year to market, about 55,000 bales of cotton.—*Gaz.*

Southern Patriot, July 17, 1827

Unfortunately the Columbia Canal depended on water from the dam at the mouth of the Broad River, and that dam had been poorly constructed. Because it failed to keep sufficient water in the canal, shipping was limited. Finally, in 1836, the legislature appropriated forty thousand dollars to repair the dam and extend the canal. The canal was reopened in January 1840, but on May 6, 1840, the dam was severely damaged by one of the largest floods in the history of the valley. The state legislature refused to appropriate additional funds for the dam's repair, and in 1843 the canal was sold for use as a power canal.[102]

The State Road

The State Road has sometimes been characterized as a disaster, primarily as a result of poor performance by its first contractor.[103] After these problems were corrected, available evidence indicates that the State Road from Columbia to Charleston was a success for transporting cotton to Charleston. For example tolls collected on the

State Road at Dean Swamp, "considerably exceeded the interest of the work, and it has been rented for the next year, for about 15 per cent on the cost of the construction of the Bridge and Causeway."[104] Another example of the success of road transportation is provided by the experience of the Columbia company established to build bridges over rivers to connect with the State Road. By 1827 this company had successfully constructed a bridge 1,400 feet long on 32-foot-high granite piers across the Congaree River at Columbia. The company reported that in the first year of operation tolls for the bridge yielded a 7 percent return on an eighty thousand dollar investment. Because of this success, they planned two additional bridges: one across the Saluda River two miles west of Columbia, which was to be 450 feet long on granite piers 28 feet high at a cost of thirty thousand dollars, and a bridge over the Broad River three miles above Columbia, which was to be 800 feet long on piers 24 feet high at a cost of thirty-six thousand dollars. The Saluda River bridge was to include five miles of turnpike.[105]

The result of improved navigation for steamboats and canal boats and the development of the State Road was a significant increase in the amount of cotton shipped from the upcountry to Charleston. Cotton shipments almost doubled in the period between 1820 and 1827:

The following interesting comparative statement has been furnished us by a gentleman, who has framed it with care from data, as correct as it is possible to obtain. We believe its substantial accuracy may be relied upon.

Statement of the Exports of Cotton and Rice from Charleston,
FOR THE LAST EIGHT YEARS.

	Bales of Upland Cotton.	Bales of S. Isla'd Cotton.	Tierces of Rice.
From 31st September 1819 to 1st Octob. 1820.	125,475	21,484	64,153
" " '20 " " '21	98,678	24,682	73,366
" " '21 " " '22	104.540	25,510	78,161
" " '22 " " '23	136,166	26,747	80,398
" " '23 " " '24	129,886	24,632	102,170
" " '24 " " '25	141,074	18,253	92,577
" " '25 " " '26	164.543	12,647	95,193
" " 26 " " '27	199,175	31,828	108,643

There was exported to London in 1824, 92,703 bush. rough Rice.
" " " " '25, 66,853 " " "
" " " " '26, 128,750 " " "
" " " " '27, 121.027
" " Liverpool " 48,980, 170,007 " " "

The rough Rice is included in the above statement of Exports of clean Rice, at the rate of 21 bushels to the tierce.

Southern Patriot, October 2, 1827

Problems with Internal Navigation

The South Carolina construction program was in many respects an immense achievement. Roughly two million dollars was expended in a ten-year time frame to open twenty-four hundred miles of river navigation, build numerous canals, and construct the State Road. But the canals, with the exception of the Columbia Canal, were not extensively used (see table 5.2).

The primary reason more boats did not use these navigations systems was the extreme variability of water levels from drought to flood. A second important reason was that the smaller rivers cleared for navigation did not remain cleared. Floods deposited debris in the channels; fishermen built weirs; woodcutters dumped logs into the river; and normal sedimentation filled channels. Such problems were identified by the Board of Public Works as early as 1821: "The rivers when once cleared of logs, are subject to freshes, to have others brought down or thrown in by the washing of the banks. These are generally precipitated into the boat channel, with all their branches and roots, and present serious obstructions of themselves, and if permitted to remain long, form bars, and change or close up the channel; at first they are easily removed, but when covered with rubbish, sand or gravel, their removal is a matter of considerable difficulty."[106]

Blanding identified that this problem on the Great Pee Dee River prevented boat usage: "The boat channel 150 feet wide has been cleared of logs from Cheraw to Georgetown. But although this work had been completed but about one year, when I descended this river, in May last, yet I discovered that in several places new depositions of logs had taken place, some of which were much in the way of the passage of boats. And there can be no doubt that however well this or any other river in this state may be cleared, similar obstructions will be formed, so long as the banks of our streams are covered with timber."[107]

To address this problem, Blanding recommended the appointment of commissioners who would be responsible for clearing the river: "It is important, therefore, that some plan should be devised by which every obstruction to the navigation of our rivers, which have been once cleared, should be removed as soon as they are formed. To do this by drafts on the public treasury, and through this department must be attended with great delays and a vast increase of expense. It can only be done cheaply and expeditiously by means drawn from those who use the stream, and by commissioners of the vicinity, who can act promptly, and thereby avoid an accumilation of obstructions."[108]

Blanding envisioned that the state could enter into a modest contract with boat owners to clear the rivers, saying that he "found that it was probable that a contract might be made with some of the owners of boats that constantly navigate that river, by which it might be always kept clear of obstructions. for less than $500 per annum; a boat descending the river with light machinery prepared for the purpose, would be delayed but a short time in removing recent obstructions, which are never difficult to be removed until they have become imbedded in the sand, or have been permitted greatly to accumulate."[109] In 1821 the Board of Public Works recommended a different solution to this problem, river police: "The board therefore take the liberty of recommending the propriety of establishing a police, on each river or section of a river, with power to call out some portion of the male population in its neighborhood, to remove these occasional obstructions."[110]

This measure related to the board's recommendation for using the levy to clear the rivers, an approach legislated before the War of 1812. The Board of Public Works recommendation of 1821 was something more. In addition to establishing a levy, the board wanted these river police be given powers to "prosecute all offenders, for felling trees or throwing other obstructions into the river."[111]

This river-police recommendation was renewed in the board's report of 1822 and in the superintendent of public works's reports of 1823 and 1827. The 1827 report, for example, stated: "I feel it my duty again to call the attention of the Legislature to my former recommendations of the establishment of a river police on each of the navigable streams of the state which have been opened.—Nothing short of this can relieve the public treasury from incessant demands for the removal of new obstructions, which must occasionally occur, and which could be removed at a trifling expense, were all the streams put under the charge of local commissioners."[112]

But sentiment within South Carolina had turned against additional funding for canals and waterways. The legislature was not about to establish a river police, nor were they about to appropriate more money for removing obstacles in rivers already done. The river-navigation schemes were not working, and if the river channels above the Santee Canal were blocked, then the traffic through the Santee Canal would suffer.

Beginning in 1823, the legislature refused to spend additional funds to improve navigation on rivers that had already been improved. By 1828, with a downturn in the price of cotton, the legislature halted all expenditures on internal improvements. Without maintenance the rivers above the Santee Canal became less safe and less navigable. As the navigation on rivers deteriorated, alternative transportation means were established. The State Road between Columbia and Charleston carried more freight. Then the railroad came.

Development of the Railroad

Although traffic on the canals and waterways "above the falls" never materialized to any extent, traffic "below the falls" was substantial and, with the help of the State Road, it doubled the flow of cotton into Charleston by the end of the 1820s. Seeking additional means to increase cotton flow to their port city, Charleston merchants looked beyond Columbia and central South Carolina to Augusta and Hamburg on the Savannah River. If a canal or a railroad could be built to the upper reaches of the Savannah River, they estimated that another one hundred thousand bales of cotton could be diverted from Charleston's rival port city, Savannah.[113] Transporting salt, shoes, clothing, and other merchandise up the Savannah River, they believed, would open the Georgia market to merchandise from Charleston.

The immediate cause for the Charleston merchants to call for the opening of an alternative route to Augusta and Hamburg was the low water level of the Savannah River in 1827 and the resulting suspension of steamboat operations. Initially the merchants called for the construction of a new canal between the Ashley River and the Savannah River. By the end of 1827 letters began to appear in the Charleston newspapers calling for the legislature to develop such a canal: "The interruption to the intercourse between Charleston and the interior of the State—the stagnation of the sales of produce, and the consequent suspension of business, generally at the briskest period of the year—presents a state of things in the Metropolis that loudly demands inquiry and interference on the part of the Legislature. With this end some views were recently presented through the papers of this place, as to the practicability of opening such a communication between Ashley & Savannah river, as would insure the arrival of our usual supplies of produce."[114]

Such a communication between the two rivers would require a canal from the Ashley River to the Edisto River and a second canal from the Edisto River to the Savannah River. In the view of its proponents, such a canal would require "less than forty miles of easy canaling."[115] But they were aware that the State of South Carolina had already turned down a canal from the Ashley River to the Edisto River and that sentiment in the legislature was against additional, expensive internal improvements: "One thing is certain; it [construction of a canal from the Ashley to the Edisto] will not be done by the State. Her resources are crippled, and her ardour cooled, by what she has done already [in Internal Improvements]. If it is to be done at all, at least for many years to come, it must be accomplished by means of private capital."[116]

A possible alternative to this canal was a railroad. No railroad had yet been built in America. One railroad advocate was architect-engineer Robert Mills, who had previously called for the construction of a grand canal between Charleston and Columbia. According to Mills, "The great superiority which Railroad possess over Canals, in every point of view, whether as regards economy, facility of communication (and that at all seasons of the year and all states of the weather—rapidity of transportation—adaptation to every circumstance of country whether a level plane or composed of hill and dale—whether crossed by mountains or broken by foaming torrents—the ease with which obstacles of this nature can be overcome, give every assurance, that for great commercial purposes, *Rail-roads must take precedence of Canals*" (emphasis in original).[117]

A public meeting was called for December 6, 1827, to petition the legislature on the subject:

> **PUBLIC MEETING.**
> Upon the application of many citizens, the Intendant ordered an extra meeting of Council for yesterday afternoon, when it was unanimously resolved to call a meeting of the Citizens TO-MORROW, at one o'clock, P. M. at the City Hall, to petition the Legislature to cause the country between Savannah and Ashley Rivers, to be surveyed, with a view to a canal that will connect them, or a Rail Road from Augusta to Charleston.

Southern Patriot, December 5, 1827

Two days later, on December 8, 1827, Charleston representative Alexander Black entered a bill in the South Carolina House of Representatives to incorporate a company to construct a railroad between Charleston and Hamburg, Columbia, and Camden.[118] Prior to its second reading, the bill was referred to a legislative committee composed of Alexander Black, B. F. Duncan, B. F. Whitner, A. P. Butler,

Thomas Lang, and C. Boohter.[119] Five days later, on December 13, 1827, the committee reported out a bill, "To authorize the formation of a Company for constructing Rail Roads or Canals, from the city of Charleston to the towns of Columbia and Camden, and to the Savannah river at or near the town of Hamburg."[120] This new company was to be called, "The South Carolina Canal and Rail Road Company."[121] The legislature passed this bill into law.

On February 18, 1828, a public meeting was held in Charleston to appoint a committee "to collect subscriptions and cause the country between Charleston and Hamburg, to be surveyed for the establishment of a Rail Road." Dr. Joseph Johnson was named chair of the committee, and Edwin P. Starr was appointed secretary. Timothy Ford, Thomas Bennett, Thomas Napier, William Aiken, and Alexander Black were named to the committee.[122]

The development of the Charleston to Hamburg railroad continued to move quickly. On March 13, 1828, the committee published its estimate of what such a railroad might cost—thirty-five hundred dollars per mile. This estimate was well received, and newspaper accounts speculated that the actual cost might be even less, perhaps three thousand dollars per mile.[123] Subscriptions for stock in the new railroad company opened on March 17.[124] Two days later, the Charleston newspapers were reporting that five-sevenths of the capital needed to organize the Canal and Rail Road Company had already been subscribed, including twenty-thousand dollars from the Charleston City Council.[125]

The first stockholders' meeting of the new South Carolina Canal and Rail Road Company was held on April 28, 1828, chaired by Timothy Ford. A committee of seven was appointed to draft the company's bylaws: Hon. John Gardner, Colonel Abram Blanding, Alexander Black, Thomas Napier, L. Holmes, T. Tupper, and Timothy Ford.[126]

Plans for the new company continued to move quickly. By August 8, 1828, the new company was asking for bids on a detailed survey of a portion of the proposed route for the new railroad:

Rail Road Company.

SEALED Proposals will be received by the Secretary of the Company, for "a plan or map, on a scale of two inches or more to the mile, of the country, along the eastern margin of Ashley River, from the mouth of Eagle Creek, across in an easterly direction, three miles, at or near Ladson's, thence in nearly a straight line to the juncture of the Goose Creek and Dorchester Roads, thence taking the road to the Meeting-street fork, and along Meeting-street, to where the lines intersect. This plan to be formed by connecting the surveys of the several plantations, transferring the character of the surface from the private plans, exhibiting the location of buildings, and names of the owners."

Separate Proposals will be received for "furnishing a plan of the country, three miles west of a direct line from the Charleston Bridge, to the Cross Roads nigh Bacon's Bridge—on a scale, and in a manner similar to the above."

Proposals must be handed in on or before the 30th instant, EDWIN P. STARR,
August 1. 13 *Secretary.*

Southern Patriot, August 8, 1828

The new company also requested assistance from the United States War Department, "for Engineers to survey the country between this city and Hamburg, with a view to the location of a Rail Road."[127] The response from the "Engineer Department to the President" indicated that the survey would commence in November 1828.[128]

Although this survey had not yet been undertaken, a stockholders' meeting was held in August 1828 to consider "the propriety of immediately ordering the Iron necessary for the contemplated railroad." At the same time, the U.S. Treasury instructed the Charleston collector of duties to levy only a 25 percent duty on iron to be imported for the purpose of constructing the railroad, rather than the usual duty of thirty-six dollars per ton, thus saving the company thousands of dollars.[129]

So. Ca. Canal and Rail Road Company.

NOTICE is hereby given, that a meeting of the Stockholders in the above Company will be held in the hall over the Office of the Fire and Marine Insurance Company, Broad-st. at 12 o'clock M. *Thursday*, 4th of September next, to take into consideration the propriety of immediately ordering the *Iron* necessary for the contemplated Rail Road &c.

By order of the Board of Directors.

EDWIN P. STARR, *Secretary.*

Rail Road Office
Charleston, August 20th, 1828. ws4

Southern Patriot, August 30, 1828

There were several surveys done of the proposed route from Charleston to Hamburg. The first was undertaken from February to June 1828 by Charles Parker and Robert K. Payne.[130] This route was resurveyed in 1829 by Horatio Allen,[131] recently appointed as chief engineer for the new railroad company.

In 1830 Allen and the railroad company engaged in experiments using wind power as the motive power for the new railroad. On March 20, 1830, they tried using a railroad car equipped with a five-by-nine-foot sail. Their experience was described in the *Charleston Courier:*

> Fifteen gentlemen got on board and flew oft at the rate of twelve to fourteen miles an hour. Thirteen persons and three tons of iron were carried about ten miles an hour. The preparations for sailing were very hastily got up, and of course were not of the best kind, but owing to this circumstance, the experiment afforded high sport. The wind blew very fresh from about north east which, as a sailor would say, was "abeam," and would drive the car either way with equal speed. When going at the rate of about twelve miles an hour and loaded with fifteen passengers, the mast went by the board with the sail and rigging attached, carrying with them several of the crew. The wreck was described by several friendly shipmasters who kindly rendered assistance in rigging a jury mast, and the car was again soon put under way. During the afternoon the wind changed so as to bring it nearly ahead when going in any direction; but this did not stop the sport as it was ascertained that the car would sail within four points of the winch, We understand it is intended by some of the seamen to rig a car properly and shortly to exhibit their skill in managing a vessel on land.[132]

Instead of continuing with wind-powered cars, Allen turned to steam power. In 1830 he contracted with the West Point Foundry in New York City to build what would become the first railroad steam engine sold in the United States—the *Best Friend of Charleston.*[133]

With the new engine, the South Carolina Canal and Rail Road Company became the first regularly scheduled railroad in the United States.[134] The American railroad revolution had begun. Steamboat and canal boat traffic declined. Although water transport continued to have a cost advantage over the railroad, planters preferred to ship their cotton by rail. Water transport was subject to disruption by low water, floods, and river debris. Cotton planters, like most farmers, borrowed money to finance their operations. Delays in bringing their cotton to market meant increased interest costs. Although railroad shipping tended to be more expensive than water transport, it was quicker and less likely to be disrupted by drought or flood. Increasingly cotton planters turned to railroads to transport their cotton. South Carolina canals began to close, and cleared waterways became blocked by trees, sediment, and other obstructions. The canal and waterway era was over.

Closing of the Santee Canal, 1853

By 1840 all the river-navigation systems and canals that fed traffic to the Santee Canal, except the Lockhart Canal on the Broad River, were out of business.[135] At its height of prosperity, say 1827–30, the Santee Canal Company was paying annual dividends of $20 per year.[136] From 1841 to 1849 dividends of the company averaged only $1.28. By 1841 the Santee Canal Company had subdivided its shares. Each $300 share, once worth $800, was subdivided into parcels of $25 each. By 1849 operating expenses were exceeding income, and in 1853 the board of directors petitioned the General Assembly to surrender its charter.[137] There was opposition to this, and it was proposed to dig the Santee Canal deeper.[138] Nothing came of this proposal and the company went out of business in 1853.[139] The lower reaches, from lock 8 to the Cooper River, were used by local planters until 1865.[140] The use the state's rivers as highways seemed dead. However, following the Civil War, there was one more attempt to revive the waterway navigation.

6 ~ Aftermath

> Cotton, the principal staple, is in most cases raised on borrowed capital. . . . This capital is furnished in the form of provisions and fertilizers on which the producer has been obliged to pay, heretofore, an average of 2½ per cent. interest per month. It is therefore of great importance to the producer to send his crops to market by the most expeditious mode he can find, and the railroad is for that reason generally preferred by him to the slower river or canal route.
>
> U.S. Senate, Ex. Doc. no. 60, *Letter from the Secretary of War,* February 28, 1881

By the time the Civil War broke out in 1861, South Carolina's canals and river-navigation systems had been abandoned. Nonetheless the idea of using South Carolina's waterways for transportation was not dead. Following the Civil War, fueled by indignation at the high cost of railroad transportation, efforts were made to resurrect water transport in South Carolina, particularly the Santee Canal.

Resurrecting the Santee Canal

After the Santee Canal was closed in 1853, it was still used by local traffic. In 1880 and again in 1912, efforts were made to resurrect it.

On June 14, 1880, the U.S. Congress passed the River and Harbor Act of 1880 to undertake the survey of improvements to rivers and harbors. A portion of this was intended for assessing the economic feasibility of bringing the Santee Canal back into operation. This study was undertaken in late 1880 by Captain James C. Post, who was assisted by J. H. Fisher and under the supervision of Lt. Colonel Q. A. Gillmore.[1] The study estimated that the Santee Canal could be brought back to good order for an estimated $100,000—which did not include the cost for providing adequate water to the Santee Canal during drought conditions:

To dredging 252,683 cubic yards of mud, at 25 cents	$63,170 75
To clearing 723,007 square yards, at $140 per 1,000 square yards	1,012 20
To rebuilding 22 gates, at $240	4,400 00
To rebuilding tide-lock	2,500 00
To repairing waste-ways and flood-gates	3,500 00
To repairing 21 bridges, at $25	525 00
To repairing stone and brick work on 10 locks, at $172	1,720 00
Total	$76,827 95[2]

The army rounded this number up to $100,000 to include engineering, contingencies and other expenses.

The study also discussed the problem of providing water to the canal during drought conditions. Captain Post suggested three alternatives:

1. Lower the summit level of the canal.
2. Use steam pumps to draw from the Santee River.
3. Restore the old reservoirs and augment them with steam-powered pumps during the dry season.[3]

The third alternative was discarded as not offering any advantage over the second alternative, which would supply sufficient water at the summit level with pumps.

To lower the canal by thirty to forty feet (the first alternative), Captain Post recommended that the present canal section at the summit level be excavated an additional thirty to forty feet. This would obviate the need for locks, except a guard lock, between the Santee River and lock 9. This option would require an excavation of approximately five million cubic yards of what was believed to be soft, alluvial soil at an estimated cost of $891,000.

5,000,000 cubic yards excavated, including revetting, at 15 cents	$750,000
For locks, basins, buildings, bridges, &c.	60,000
	810,000
Add 10 per cent for engineering and incidental expenses	81,000
Total	891,000[4]

Using steam-powered pumps, the second alternative, would be less expensive. Captain Post estimated three of these pumps would be required, each with an approximate capacity of ten thousand gallons per minute pumping water through four hundred feet of fifteen-inch diameter pipe. The total cost would be approximately $45,000:

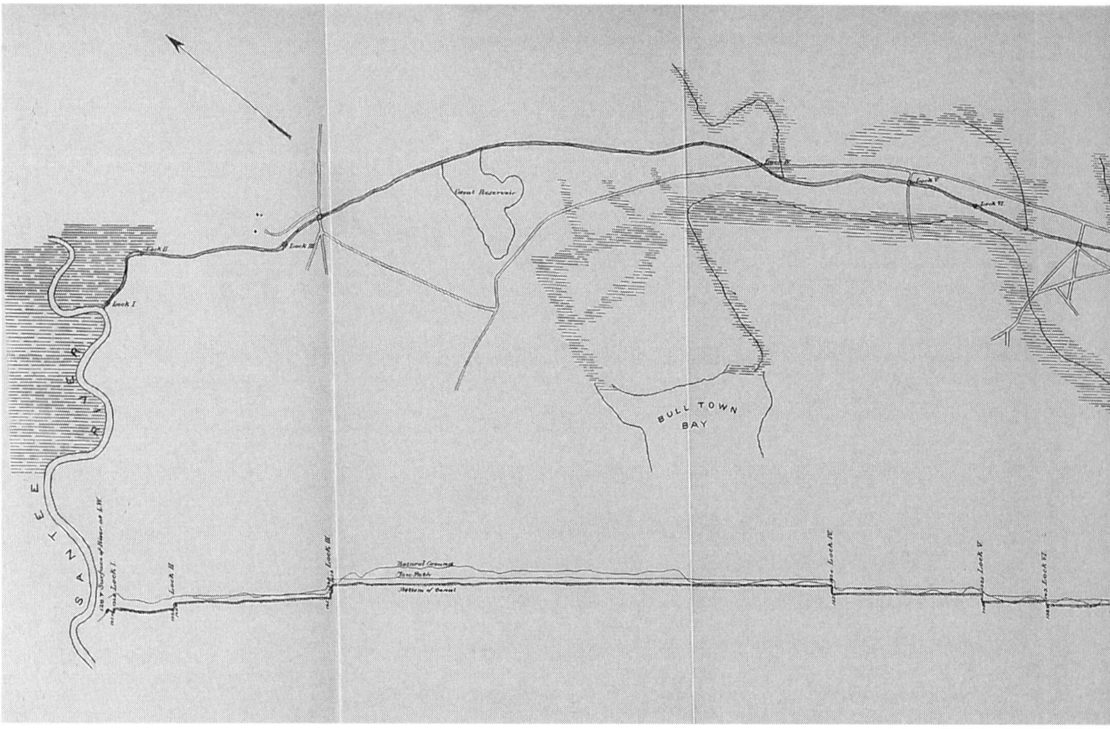

Fig. 6.1. *Above and facing:* Captain James A. Post's 1880 survey of the Santee Canal. From U.S. Senate, Ex. Doc. no. 60, *Letter from the Secretary of War* (1881)

Pump No. 1, with engine and boiler	$5,000
Pump No. 2, with engine and boiler	7,300
Pump No. 3, with engine and boiler	10,500
400 feet 15-inch pipe, at $7	2,800
12 elbows, at $50	600
Pipe guards and other fixtures	1,000
Building	13,500
Add 10 per cent. for contingent expenses	4,070
Total	44,770[5]

This estimate, however, was only a first-time, or capital, expense. To the above amount, Captain Post added capitalized annual expenses—what we would call today life-cycle costs—for this second alternative.

1 chief engineer, $125 per month	$1,500
3 assistant engineers, each $75 per month	2,700
3 firemen, each $50 per month	1,800
Oil, tallow, waste, and repairs	2,700
1,800 tons of coal, at $6.50	11,700
Total	20,400[6]

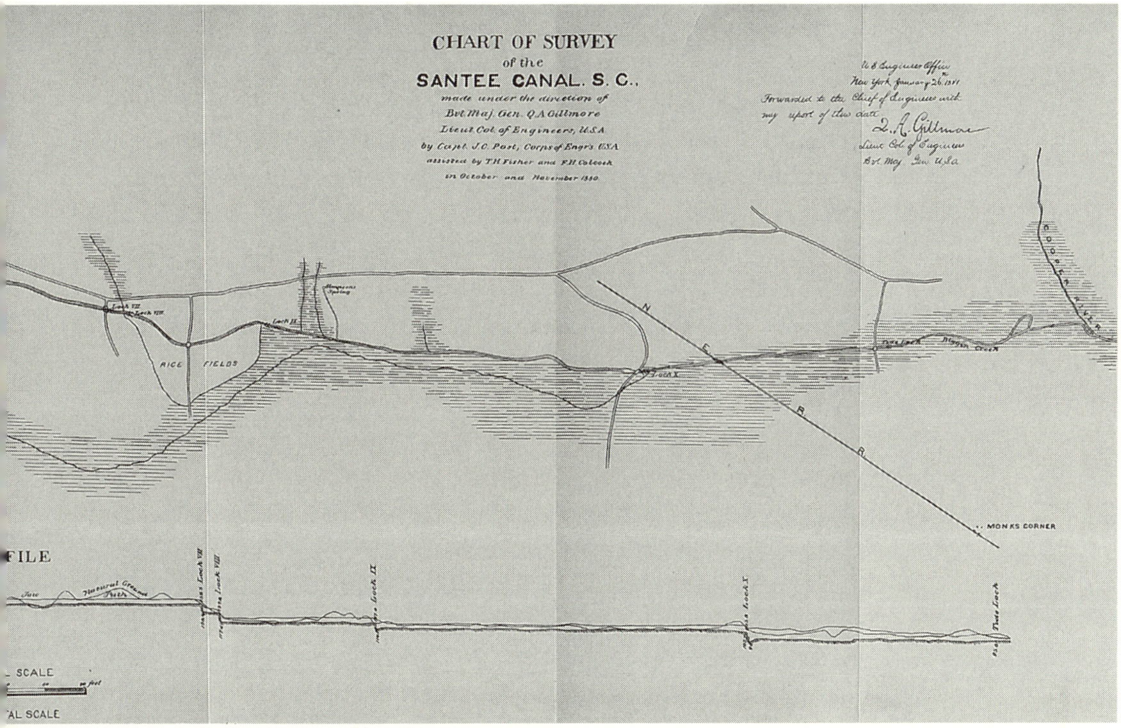

Capitalizing $20,400 at 6 percent interest would be equal to $340,000. The total capitalized cost of alternative 2, using steam-powered pumps, would be:

Repairing canal and locks	$100,000
Procuring and setting up pumping plant	44,700
Running expenses of pumping plant, capitalized	340,000
Total	484,770[7]

Captain Post also provided the annual operating expenses of the Santee Canal (not including the cost of pumping), using Colonel Senf's report of 1800, but using 1880 salaries:

2 toll receivers, at $900 per annum	$1,800
10 lock-keepers, at $500 per annum	5,000
10 laborers, as assistants at locks, at $300 per annum	3,000
2 blacksmiths, at $500 per annum	1,000
4 carpenters and boat-builders, at $500 per annum	2,000
2 bricklayers, at $500 per annum	1,000
	13,800
40 per cent. for superintendence, tools, materials, and incidental exp.	5,520
Total	19,320[8]

Therefore the water-pumping alternative, alternative 2, would be the least expensive of the options examined by the army.

It should be noted that the army did not consider increasing the width or depth of the canal, important aspects of the Board of Public Works examinations in 1823. That improvement would have added cost to either alternative. In the end the army recommended against rehabilitating the Santee Canal. In particular they cited the basic problem of river navigation—that obstructions were allowed to accumulate in the river to block traffic: "The transportation business of that river [the Congaree and the Broad] has much diminished since the railroads were built that run near and parallel to it, or through the districts that supplied it with freight. At the same time it may be said that an improvement of this river and other rivers in South Carolina is urgently called for by many people interested in the transportation business, and that the ascendancy of the railroads is at least in part ascribed to the many obstructions that are allowed to impede river navigation, the removal of which would attract business to these rivers."[9]

The army thus tied the success of the Santee Canal to success of river navigation above the canal. And the success of that river navigation was tied to the need for clearing the obstructions in the river—the very point that the South Carolina legislature decided not to address when the establishment of a "river police" was recommended to them by the Board of Public Works.

Finally the army report pointed out the one advantage that railroads have over canals—speed, for which growers were willing to pay a premium in order to shorten the term of their crop loans. Railroads had an inherent advantage over the slower canals.[10] Interruption of water transport by droughts and floods provided an additional advantage to the railroads. No improvement to the Santee Canal would have eliminated these advantages. No further action was taken on the canal at that time.

In 1902 South Carolina congressman William Elliott introduced a joint resolution calling on the secretary of war to make another survey for joining the Santee and Cooper rivers by canal just north of the old Santee Canal.[11] The army took no action on this resolution. In 1909 Congress passed another River and Harbor Act to conduct a nationwide survey of inland navigation.[12] Although it is not clear, this act may have been the reason for the 1912 maps prepared at that time for the Santee Canal.[13]

The Santee Canal was the subject of one final U.S. government study. In 1934 the General Assembly of South Carolina passed Act 887 establishing the Santee Cooper hydroelectric and inland navigation. It was clear that most of the remains of the old Santee Canal would be inundated by the impoundment (called Lake Moultrie) by Pinopolis Dam, constructed in the vicinity of lock 10, Simpson's Lock. Thomas Waterman of the National Park Service's Historic American Buildings Survey (HABS) conducted a survey to photograph the plantation houses that would be lost to the impoundment.[14] These photographs are now included in the HABS collection in the Library of Congress. Not included in this study, but undoubtedly photographed at the same time by HABS photographers, are the four HABS photographs of the Santee Canal locks included in this book.

Retrospective

At the end of the eighteenth and beginning of the nineteenth centuries, the immense western lands of America offered the promise of great wealth, but a means had to be found of transporting the agricultural plenty of these lands to seaports such as Charleston for export to the Caribbean or Europe, as well as for shipment to U.S. cities. South Carolina's internal-improvement program was one of America's answers on how to ship the wealth of the western lands to the coastal cities.

Like New York, Pennsylvania, Maryland, Virginia, and other states, South Carolina turned to water for its transportation needs. With their state's proliferation of rivers and dearth of roads, South Carolinians saw shipping by water as the best way to carry the agricultural wealth of the upcountry to the coast, especially since existing river passages could be economically tied together with new canals. At only one half the cost of land transport, shipment by water was very attractive.

One of the first American canal efforts, the Santee Canal was built when few Americans had even seen a canal lock. At a time when there were very few American engineers, South Carolinians, like their northern cousins, turned to Europe to find engineering expertise. While the Santee Canal was designed and built by a German, John Christian Senf, most of the early northern canals were designed and built by Englishmen such as William Weston, James Brindley and Benjamin Henry Latrobe. Later, when South Carolina undertook a massive canal-construction and waterway-improvement scheme, Americans designed and supervised the work. In most cases, these canal builders—including Abram Blanding—were self-taught. Their acquired expertise was later used in a host of other enterprises, such as railroads, road building, and banking.

The Santee Canal enjoyed modest success, but its utility was continually limited by limitations to the navigation on the rivers that fed into the Santee River from the north. Nonetheless, when canal mania swept the United States after the end of the War of 1812 and the successful completion of the Erie Canal in 1825, South Carolina undertook an ambitious program to clear its rivers and build bypass canals around river obstructions. The Santee Canal was the critical link in this water-based transportation system.

Although it enjoyed some initial success, this ambitious canal and river-navigation program did not prevail in the long term. Cleared rivers did not remain cleared. After having spent almost two million dollars on these waterways and related works, the state legislature was in no mood to spend additional funds. Navigation on these rivers conflicted with other uses, such as water-powered mills and fishing, and South Carolina's navigation-improvement effort started at the time when the severe droughts of 1816 and 1818–19 rendered the Santee Canal useless and impeded navigation on the rivers.

Nonetheless the South Carolina waterway projects were an important step in the continuum of improving transportation to the west. As New York's Erie Canal led to the New York Central Railroad, as Pennsylvania's Mainline Canal led to the Pennsylvania Railroad, and as Maryland's Chesapeake and Ohio Canal led to the Baltimore and Ohio Railroad, so South Carolina's program led to the development of the Southern Railroad and other railways in the region. Road construction was stimulated as well.

In looking back at the construction of the Santee Canal and other related canal projects, one is struck by their extremely ambitious nature. Not only were these projects enormously expensive, but they were undertaken at a time when few Americans had ever seen anything like them. The Santee Canal was particularly ambitious as it was the country's first summit-level canal—that is, the first American canal to be built from one watershed to another. This was the most difficult type of canal to build because of the difficulty of providing adequate water at the summit level.

For the time that South Carolina's canals and improved waterways were in operation, the scope of the system was impressive. The state had opened approximately 2,400 miles of waterways to navigation, insuring that most parts of the state could be reached by boat and thus having a beneficial impact on the state's economy. The amount of cotton shipped by water from the upcountry to Charleston doubled from 1820 to 1826, from almost one hundred thousand bales to two hundred thousand.[15] The improved waterways were also responsible for decreasing the cost of shipping cotton from Columbia to Charleston by half, from $2.50 per bale to $1.25 per bale.[16] The same waterways also were responsible in lowering the costs of commodities shipped from the coast to the upcountry. As the Board of Public Works reported, "It is then demonstrable that the diminution in the price of salt alone, occasioned by rendering navigable our rivers, will annually reimburse all that the upper country pays into the treasury, and more than their proportion of the expense of the public works."[17]

Eventually the canals of South Carolina were overtaken by new technologies, succumbing to competition from railroads and macadamized roads. In the end the Santee Canal and its related canal projects may be seen in a continuum of costly experimental projects devised to solve the problem of hauling America's inland agricultural wealth to the coastal ports. Nonetheless these canals and waterways were magnificent in their conception and execution. Their few physical remains are a historical monument to the spirit, determination, and innovation of the men who formulated and implemented this system of transportation.

Appendix

Rules of the Santee Canal Company

The following rules were agreed to and adopted by the present stockholders in the Santee Canal Company, at a meeting held on Monday the 10th day of January 1792:

1. THAT the stock of the company shall be divided into, and consist of, one thousand shares.
2. That on the third Monday in January in every year, the company shall meet in Charleston, when a president and twelve directors, and a secretary shall be chosen by ballot from among the proprietors.
3. That in voting, every proprietor shall be entitled to one vote for every share by him or her held, as far as five shares inclusive, and one vote for every five shares exceeding the first five shares: Provided, that no proprietor shall in any case be entitled to more than ten votes; and any proprietor, by a writing under his hand, executed before one witness, may depute any other proprietor to vote as proxy for him or her, at any meeting.
4. That no persons shall be elected to the office of president, director, secretary, or any other office, unless he shall have a majority of the votes present.
5. That no person who shall become a proprietor by transfer shall be entitled to vote for a president, director or any other officer, unless his or her right in the stock of the company shall have been registered in the books of the company three months previous to his or her voting.
6. The president and six directors shall form a quorum, and shall transact all such business as shall be committed to them by the said company.
7. In case of the absence from the state, incapacity or death of the president, the directors or a majority of them shall choose from among themselves a person to discharge the duties of that office, until the next general meeting and election; and in case of a vacancy in the directorship, the directors or a majority of them shall fill up such vacancy by ballot, to continue until the next general meeting and election.
8. That all orders and transactions by the president and board of directors, be copied in a book or books to be kept for that purpose, and signed by the

Published in the City Gazette and Daily Advertiser *(Charleston, S.C.), February 1, 1792, 2*

president—that an account of all their proceedings be laid before the company at their meetings; and that on finding them fairly and justly stated, the proprietors then present or a majority of them shall give a certificate thereof, a duplicate of which shall be entered in the books.

9. Every proprietor may transfer his or her share in the stock of the company, by deed executed before two witnesses and registered, after proof of the execution thereof, in the company's books, but not otherwise, except by devise, which devise shall also be exhibited to the board of directors, and registered aforesaid, before the devisee shall be entitled to draw any part of the profit arising therefrom: Provided, that no transfer whatever shall be made, except for one or more shares, and not for part of such shares.

10. That where a person has paid any part of the sum required by any former resolution of the company, such person shall be entitled to all the rights and privileges of a proprietor, upon his paying, on or before the first day of April, into the hands of the hon. David Ramsay, Thomas Jones and Nathaniel Russell, Esquires, all his arrears, notwithstanding his share may have been sold by order of the company, or notwithstanding he may have declined voluntarily by himself or agent.

11. That the said David Ramsay, Thomas Jones and Nathaniel Russell, shall open books on the first day of May next, at ten o'clock in the forenoon, and keep them open until two o'clock in the afternoon of the same day, and so from day to day until the fifth of the said month inclusive, for the purpose of receiving subscriptions, for seven hundred and sixty-seven shares, unless the said shares shall be subscribed on some day previous to the fifth. And if at the shutting of the books on any of the days the number of shares subscribed shall exceed seven hundred and sixty-seven, in such case the deductions shall be made in due proportion from the numbers subscribed on that day. Provided, that the present proprietors shall be at liberty to double their present shares without leaving the same liable to deduction, and with out lessening their right of subscribing in common with other persons for other shares.

12. That at the time of subscribing, the sum of five pounds be paid by each subscriber, for every share which he or she shall subscribe, into the hands of the David Ramsay, Thomas Jones and Nathaniel Russell, Esquires, which money shall be deposited by them in the bank for the use of the company.

13. That every proprietor shall pay five pounds on each share, on the third Monday in January, 1793; and the like sum on the third Monday in July 1793; and on the third Mondays in January and July, in every succeeding year, until the canal shall be completed: But if the president and a majority of the directors shall judge it necessary to raise a larger sum, they shall be at liberty to call for the same, to be paid on notice of six months; provided that the sum to

be called for on each share, shall not exceed five pounds; and provided also, that the requisitions shall be made on the third Monday in January or July.

14. That if any proprietor shall fail to make of any payments, to which by the rules of the company he or she may be liable, when it shall become due, or within three days thereafter, every such proprietor shall forfeit to the company all the preceding payments, and also all his or her interest in the said company.
15. That no person, company or body politic, or corporate, shall be allowed to subscribe for more than ten shares in any one day.
16. That none of the foregoing rules shall be altered at any meeting, unless there shall be present two thirds of the shares subscribed for: nor then, unless two thirds of the shares voting shall agree to the alteration.

NOTES

CHAPTER 1. *Canals Come to South Carolina*

1. *Report of the Board of Public Works, to the Legislature of South Carolina for the Year 1822;* reprinted in Kohn, *Internal Improvement,* 169.

2. A contemporary account of the impact of the Great Flood of 1796 on bridges near Columbia is as follows:

> The incessant rains which fell during the last week, have produced a scene of destruction, which every feeling heart must contemplate with pain. The freshes in all the adjacent rivers have been by far the greatest ever known, and consequently the most destructive.
>
> Several bridges on Saluda river have been carried off; among which is that belonging to Mr. Kennerly, who has also sustained considerable loss in houses, fences, &c.
>
> To the catalogue of losses sustained by this destructive deluge, we are sorry to add that of Col. Hampton's bridge, lately erected on the Congaree river, at the village of Granby, of which scarce a vestige remains.
>
> Several arches of a new bridge, about to be erected by Major Compty, on Broad river, are also carried off. (*City Gazette* [Charleston, S.C.], January 9, 1796, 2)

3. A typical notice announcing the use of the levy is one issued by the commissioners of Lynch's Causeway in 1826: "The Inhabitants and owners of property on North and South Santee, and in the Parish of Winyaw, who are within ten miles of Lynch's Causeway, and are therefore liable to work thereon; are informed that a portion of their Negroes is called out to put the same in order. The work will commence on Monday, the 31st inst. All persons concerned are required to make returns to either of us, of all their male slaves, from the age of 16 to 50 years, by that time" (*Southern Patriot* [Charleston, S.C.], July 17, 1826, 3).

4. Robert Mills (1781–1855), engineer, architect, and author, was born in Charleston, South Carolina and educated at the College of Charleston. He studied briefly under James Hoban, architect of the first State House in Columbia and the White House in Washington. At age nineteen Mills moved to Washington, where he renewed his association with Hoban and was introduced to Thomas Jefferson. Mills and Jefferson became friends, and Jefferson recommended him to Benjamin Henry Latrobe, then official architect of national buildings and perhaps the most experienced

architect-engineer in the United States at the time. After working for Latrobe from 1803 to 1808, Mills moved to Baltimore to establish his practice. In 1820 he moved back to South Carolina to become state architect and engineer and a commissioner on the Board of Public Works. In this capacity he was responsible for designing many courthouses throughout South Carolina. He compiled the first atlas of South Carolina, a collection of county maps. For the significance of that atlas, see Ristow, "Robert Mills' Atlas of South Carolina." Mills returned to Washington in 1830, and in 1836 President Andrew Jackson appointed him architect of public buildings, a position he held until 1852. Mills was an unusual architect-engineer in that he published a number of pamphlets and books promoting internal improvement, including *A Treatise on Inland Navigation* (1820), *Inland Navigation. Plan for a Great Canal between Charleston and Columbia* (1821), and *Statistics of South Carolina* (1826). He is perhaps best remembered as the designer of the Washington Monument. See Kohn, *Internal Improvement,* 595; Placzek, *Macmillan Encyclopedia of Architects;* Bryan, *Robert Mills: America's First Architect;* and Gallagher, *Robert Mills: Architect of the Washington Monument.*

5. Mills, *Inland Navigation.*

6. Edgar, *South Carolina: A History,* 266–67.

7. Abram Blanding (1776–1839), lawyer, commissioner of public works, and bank president, was born in Rehobeth, Massachusetts. He was educated at Rhode Island College (Brown University) under Dr. Jonathan Maxcy, who later became the first president of the University of South Carolina. It has been suggested that Dr. Maxcy encouraged Blanding to come to South Carolina. Blanding arrived in Columbia in 1797 and became a schoolteacher. In 1799 he moved to Camden and studied law under Judge Joseph Brevard. Blanding was admitted to the bar in 1802. In 1815 he married Mary Caroline DeSaussure, daughter of Chancellor Henry William DeSaussure. Returning to Columbia in 1819, Blanding constructed a waterworks (later sold to the city in 1835). In April 1820, after Major John Wilson refused the position, Blanding was appointed a commissioner on the Board of Public Works. When the Board of Public Works was superseded by a superintendent of public works in 1823, Blanding was appointed to that position, which he held throughout the duration of the work in internal improvement. After this service Blanding became the president of the Commercial Bank in Columbia and president of the Southwestern Railroad Bank. He also served on the board of the Cincinnati and Charleston Railroad Company. He died of yellow fever on Sullivan's Island in September 1839. See Kohn, *Internal Improvement,* 599.

8. Blanding, *Report of the Superintendent on the Edisto Canal,* 8–9.

9. Ibid., 9.

10. In 1791 South Carolina passed an act mandating that intestate estates were to be divided equally among the surviving children. Prior to this act, the land was transmitted to the oldest son. Blanding wrote:

> By the law of 1791, intestate estates were made to descend equally to all the children or collaterals. This law was so necessary to the preservation of our republican institutions, so congenial to the spirit of our government, and to the

> sentiments of our citizens, that wills and family provisions took generally the same equitable directions.... No one heir could take the whole plantation and remunerate the rest without disposing of his share of the slaves, which were already reduced to a number too small to cultivate the place to advantage—to divide, in many instances was to destroy it. The slaves thus were found to be the only part of the estate which, on a division were valuable. These or some part of them were removed, and the plantation was either abandoned at once or cultivated for a short time with an inadequate force, and then deserted—its banks and drains went to ruin, and rendered the preservation of the plantations adjacent to it more difficult, the culture more precarious, and the expense of the public drainage more burthensome. (*Report of the Superintendent on the Edisto Canal,* 9–10).

11. Ibid., 9. Blanding also attributed the abandonment of the rice fields of the low-country to the unhealthiness of the region:

> That all the causes combined should have produced the desolation which now presents itself throughout these inland swamps, is not surprising. But there is an evil which followed in their train, itself acting at last as a cause, that has had most powerful influence, and must not be passed unnoticed. It is an observation every day made, that the lower country has become more sickly than formerly. The fact cannot be denied as it respects this region; and I think may be easily accounted for. The rice fields once abandoned, are always covered with water, during the winter and spring, and during the whole year unless when the season is very dry. This water, from want of drainage becomes stagnant in the summer and autumn. Luxuriant vegetation springs up and decays in it. Extensive clearings have been made on the high lands for cotton fields, and thereby the exhalations from the swamps have had a free and more enlarged circulation; so that every part of this country is unhealthy. It often proves fatal to a resident of Charleston to stay one night in it during the months of July, August, September and October. The country fever as it is called, is so uniformly communicated during this season of the year to those who sleep a night within the influence of the malaria of the marshes, that it is now regarded as the greatest madness for a citizen of Charleston to visit his plantation in the country, unless he can go and return to the city in one day, without exposure to the night air. Before the abandonment of the inland rice swamps and the extensive clearing of cotton lands in this region, the country was comparatively healthy. Families resided on their plantations all the year without extraordinary disease or mortality. It would thus appear that the present unhealthiness of the country is in a great measure attributable to the neglected state of its inland swamps; and it may be foretold without much fear of the prediction being falsified, that unless some efficient measures are taken again to restore it to cultivation, this whole region will be deserted by the white population and become the resort of the banditti of the black; and this will take place in a district of country no part of which is more than thirty miles from the

populous and healthy city of Charleston. (*Report of the Superintendent on the Edisto Canal,* 11)

12. Drayton, *A View of South-Carolina,* 144.

13. Ibid., 111–12.

14. "Internal Improvement," *Southern Patriot,* December 8, 1820, 2.

15. Not all South Carolinians favored a water-transport system over an improved road system. In 1819, for example, "A River Planter," wrote a letter to a Charleston newspaper arguing the merits of turnpikes and artificial railways (that is, horse-drawn carts on tramways) over waterways. See *Southern Patriot,* January 7, 1819, 2.

16. John Wilson, engineer, emigrated from Scotland to Charleston in 1800. Like his father he was an engineer, educated at the University of Edinburgh. Wilson married Eliza Gibbes. During the War of 1812 he was placed in charge of the Charleston fortifications under General Thomas Pinckney. After the war Wilson served as a major in the U.S. Topographical Engineers and served as civil and military engineer of South Carolina in 1817–18, a position he had to relinquish when he became sick and could no longer discharge his duties. Later he was placed in charge of preparing district maps of South Carolina and "the Wilson map," the official map of South Carolina, published in 1822. In 1827, as part of Pennsylvania's internal improvements, he was appointed to make a survey for a route for a canal or railroad between Philadelphia and the Susquehanna River. See Kohn, *Internal Improvement,* 600.

17. Joel Roberts Poinsett (1779–1851), diplomat, legislator, and president of the Board of Public Works, was born in Charleston, South Carolina, and educated primarily in England and Scotland. He returned to Charleston in 1800 and traveled extensively from 1801 to 1809. In 1809, on behalf of President James Madison, Poinsett served as a confidential envoy to South America, where he negotiated several commercial treaties. Returning to South Carolina in 1816, he was elected to the state legislature. In the General Assembly, he was active in organizing the Board of Public Works and became its first president in 1820. He left that position when he was elected to the U.S. Congress in 1821. He served in Congress until 1825, when he was appointed minister to Mexico, a position he held until 1829. In 1837 President Martin van Buren appointed Poinsett as secretary of war, and he served until 1841. Throughout his career he exercised his scientific interests, introducing Euphorbia pulcherrima (Poinsettia) to the United States, presenting natural-history specimens to scientific societies, and championing the acceptance of James Smithson's bequest to establish the Smithsonian Institution. He died in 1851 in Stateburg, South Carolina. See Kohn, *Internal Improvement,* 597; and *Biographical Directory of the United States Congress,* http://bioguide.congress.gov (accessed April 30, 2009).

18. Of French birth Nicholas Herbemont taught French in 1806–7 at South Carolina College (now the University of South Carolina). An important horticulturalist of his time, Herbemont served as one of the five members of the Board of Public Works and was president of the board in 1822. He died in Columbia in 1836. See Kohn, *Internal Improvement,* 600.

19. Robert G. Mills, of Chester, South Carolina, was a member of the Board of Public Works. At different times he served in the South Carolina Senate and House of Representatives. In 1837 he became director of the Louisville, Cincinnati and Charleston Railroad Company. See Kohn, *Internal Improvement*, 600.

20. Kohn, *Internal Improvement*, xiii.

21. From 98,678 bales to 199,175 bales (*Southern Patriot*, October 2, 1827, 2).

CHAPTER 2. *Building the Santee Canal, 1786–1800*

1. "An Act for Cutting, Clearing and Making a Convenient Creek or Water-Course Through That Part of the Land Belonging to Mr. John Jones, Commonly Called the Hallover," December 18, 1714, in McCord, ed., *The Statutes at Large of South Carolina*, 7:475–76.

2. "An Act to Impower Commissioners to Make One or More Cuts or Creeks in the Parish of Christ Church, for the More Convenient Transportation of Goods Carried by Water to Charlestown, of the Persons within the Limits Therein Mentioned," June 29, 1717, in McCord, ed., *The Statutes at Large of South Carolina*, 7:477–78.

3. "An Act for Cutting and Clearing a Creek, Commonly Called Biggon Creek," March 11, 1726, in McCord, ed., *The Statutes at Large of South Carolina*, 7:481.

4. James Cook, "Letter to Mr. Timothy" [proposing South Carolina Canal], *South Carolina Gazette* (Charleston), January 18, 1770, 1–3.

5. Ibid. "Indico" is indigo. Madder is a plant that was cultivated for the reddish-purple dye derived from its root.

6. Ibid.

7. Cook wrote on this proposed canal: "It is about twelve miles from navigable water to water; upwards of eight miles and an half of which, from Biggin bridge, it is only the clearing of a creek; and from thence, by cutting about half a mile through pine land, it would meet with a creek from Santee River.—Therefore, I cannot apprehend that the expence of the above cut, will be equal to the advantages of one year; and moreover, we have many instances of much greater undertakings being performed by individuals" (ibid.).

8. There were two Henry Mouzons. This map was produced by Henry Mouzon Jr. (1741–1807), the grandson of Huguenot immigrants. It is believed that he was sent to France for his education after his father's death in 1749. In 1771 Governor Lord Charles Greville Montague appointed Mouzon and Ephraim Mitchell to survey the boundaries of the civil districts of South Carolina. Issued in 1775 as *An Accurate Map of North and South Carolina, with their Indian Frontiers*, Mouzon's map is best known for including both North Carolina and South Carolina. His cousin Henry Mouzon of Craven County, South Carolina, has been frequently confused with him. The son of Louis Mouzon, Henry Mouzon of Craven County was also a surveyor and provided many of the surveys for his cousin. See *Transactions of the Huguenot Society of South Carolina*, no. 69 (1964): 46–50; no. 70 (1965): 45–55; no. 91 (1986): 147–57.

9. This map, "A Map of the Parish of St. Stephen, in Craven County," is shown in fig. 2.1 and is reproduced in Porcher, *The History of The Santee Canal*. It was also

published in 1881 in U.S. Senate Ex. Doc. no. 60, *Letter from the Secretary of War.* The map in Porcher's history is dated 1775, but that date may have been added later. A somewhat different map of the same area, with the same title, exists in the South Caroliniana Library of the University of South Carolina, which has given it a date of 1785. In U.S. Senate, Ex. Doc. no. 60, *Letter from the Secretary of War,* the map is undated, but the report states: "According to the best information that could be obtained, this map must have been made before the war of the Revolution." See U.S. Senate, Ex. Doc. no. 60, *Letter from the Secretary of War,* 10. The map shown in this book (fig. 2.1) is a copy of the map in *Letter from the Secretary of War.*

10. "Report of Captain James C. Post, Corps of Engineers," December 13, 1880, attached to U.S. Senate, Ex. Doc. no. 60, *Letter from the Secretary of War,* 9–10.

11. *South Carolina Gazette,* November 12, 1785, quoted in Salley's appendix to Porcher, *The History of the Santee Canal.* John Faucheraud Grimké, Richard Hutson, Major Mitchell, Thomas Jones, John Dawson, Aaron Loocock, and William Bull were appointed as a committee to receive subscriptions.

12. Ibid.

13. Signers included John Rutledge, John Faucheraud Grimké, Theodore Gaillard, George Haig, James Kennedy, —— Graham, Thomas Sumter, Benjamin Waring, Thomas Walter, John Vanderhorst, James Mitchell, Aedanus Burke, Peter Tayssoux, Richard Champion, Aaron Loocock, —— Pearson, John James, Francis Marion, John Dawson, Alexander Gillon, Samuel Midwood, John Richardson, Ephraim Mitchell, William Bull, Duncan McRa, Nathaniel Russell, Philip Gadsden, Peter Belin, Henry Laurens Jr., Edward Rutledge, Ralph Izard, John Budd, Robert Beaty, William Smith, Minor Winn, William Clarkson, William Hill, James Theus, Joseph Atkinson, Thomas Jones, and Daniel Bourdeaux (dashes indicate blanks in the original). Petition for the Santee Canal Company, February 7, 1786.

14. Ibid.

15. McCord, ed., *The Statutes at Large of South Carolina,* 7:541–43.

16. Ibid., 7:541–43.

17. Directors were Judge Burke, Judge Grimké, Judge Drayton, General Thomas Pinckney, General Thomas Sumter, General Francis Marion, Commodore Gillon, Major Mitchell, Edward Rutledge, John Huger, Thomas Jones, Thomas Walter, William Doughty, Joseph Atkinson, Henry Laurens Jr., James Sinclair, Theodore Gourdin, Aaron Loocock, and Theodore Gaillard. Dan Bourdeaux was elected treasurer, and Stephen Drayton was elected secretary. *Charleston Morning Post, and Daily Advertiser,* March 25, 1786, quoted in Salley's appendix to Porcher, *The History of the Santee Canal.*

18. U.S. Senate, Ex. Doc. no. 60, *Letter from the Secretary of War,* describes Henry Mouzon as a surveyor and not an engineer. See ibid., 4.

19. James Brindley (1748–1820), American canal engineer, was probably born in Alton, England, nephew of James Brindley the famous English canal engineer. His uncle built canals near Manchester, England, for Francis Egerton, Duke of Bridgewater. It is likely that the nephew James Brindley worked for his uncle in building these canals. The earliest in the English-speaking world, these canals were studied by Englishmen and Americans, and in 1774 Brindley the nephew was recruited by John

Ballendine to immigrate to the United States and work on canals in Maryland and Virginia. Shortly after his arrival, he began work on the Little Falls bypass canal on the Potomac River immediately above what was to become Washington, D.C. The canal was never completed; the Revolutionary War intervened, and Brindley eventually enlisted in the Delaware militia. After the war Brindley worked as chief engineer on the Susquehanna Canal in Maryland, and the Conewago Canal in Pennsylvania. He consulted on the Potomac Canal, the James River Canal in Virginia, the Santee Canal in South Carolina, and others. Yvonne E. Long and Gordon Brindley, B.A., "James Brindley of Alton, Staffordshire, and the George Washington Connection," unpublished paper, December 2002, 2–3; Kapsch and Long, "James Brindley, American Canal Engineer."

20. Letter from William Moultrie to George Washington, April 7, 1786, in *The Papers of George Washington,* Confederation Series, edited by Abbott, Twohig, et al., 4:6–7 (hereafter cited as *Washington's Papers*).

21. John Ballendine (?–1781), iron manufacturer and entrepreneur, was from Prince William County, Virginia. Harrison speculates that Ballendine may have had a contract to transport troops by water to Alexandria in support of General Edward Braddock's expedition of 1755. In 1759 or 1760 Ballendine was accused of short changing George Washington on iron that Washington had paid for. In 1762 Ballendine entered into a contract with John Semple for a forge and other property on Occoquan Creek about two miles above Colchester. Ballendine later left Occoquan Creek to operate a saw mill at Seneca Falls on the Potomac River. Admired for his mechanical abilities, Ballendine was sent to England in 1773 by the governors of Maryland and Virginia and the property owners along the Potomac to study the Duke of Bridgewater's canals west of Manchester. Despite Washington's previous experience with being short changed by Ballendine, Washington was supportive of Ballendine and his mission to England. While in London, Ballendine was not successful in developing financial support for the Potomac River or James River navigations. Ballendine returned from England in 1774 with various canal workers, including James Brindley, nephew of the famous canal engineer of the same name. Ballendine began work constructing a canal at Little Falls on the Potomac, on his property called Amsterdam. He relocated from the Potomac River to Richmond in 1775 and began working on a canal there. During the Revolutionary War he constructed a cannon factory at Richmond for the State of Virginia. The factory was destroyed during Benedict Arnold's James River raid of 1781. Built by William Buckland in 1758, Rockledge, Ballendine's house at Occoquan, still stands. See Skaggs, "John Semple and the Development of the Potomac Valley, 1750–1773"; Harrison, *Landmarks of Old Prince William,* 2:435–36; Fitzpatrick, ed. *The Diaries of George Washington 1748–1799,* 1:217; and Church, "John Ballendine: Unsuccessful Entrepreneur of the Eighteenth Century."

22. See Kapsch, *The Potomac Canal;* Kapsch, "George Washington, the Potomac Canal and the Beginning of American Civil Engineering History"; and Kapsch "The Potomac Canal: A Construction History."

23. From the Revolutionary War, Washington would have known John and Edward Rutledge, General Thomas Sumter, General Francis Marion, and others. See *Washington's Papers,* 4:7n2.

24. At the time Brindley was the chief engineer of the Susquehanna Canal Company. The Susquehanna Canal was being constructed on the east bank of the Susquehanna River from above Havre de Grace to the Pennsylvania-Maryland border. For a history of this early canal see [Susquehanna Canal Co.], *Report of the Governor and Directors.*

25. Letter from George Washington to William Moultrie, May 25, 1786, in *Washington's Papers,* 4:73.

26. Letter from William Moultrie to George Washington (enclosure), August 7, 1786, in *Washington's Papers,* 4:202n1.

27. Letter from James Rumsey to George Washington, March 29, 1786, in *Washington's Papers,* 3:613n1. Brindley also consulted on the Potomac Canal, for which James Rumsey was initially principal manager.

28. Letter from James Brindley to George Washington, April 5, 1787, in *Washington's Papers,* 5:124. Brindley wrote, "I was detained in Charleston until the 17th of March and returned for expedition, by the Philadelphia Packet—therefore had not time to return by Land."

29. Senf has sometimes been identified as a Swede. Arthur P. Wade, in his history of Mount Dearborn, correctly identifies Senf as a Saxon. Wade wrote:

> Henry Laurens, a representative to the Continental Congress, had sought a military engineer among the German auxiliaries captured with Burgoyne at Saratoga. He selected John Christian Senf, a 24-year-old Saxon, and arranged—probably illegally—to have Senf appointed a captain of South Carolina troops in 1778. As Capt. Christian Senf, the young engineer served with state troops under Maj. John Faucheraud Grimké, and by the fall of Charleston he was a lieutenant colonel. After the war Colonel Senf was appointed State Engineer, in which position he was considered as available for superintending the planning and construction of the postwar canal projects, even though these involved joint-stock companies expecting a profit. When the Santee Canal Company was chartered in 1786, Senf was employed as its chief engineer.

Grimké was one of the signers of the petition for the Santee Canal on February 7, 1786. See Wade, "Mount Dearborn," 209–10.

30. Both the Santee Canal in South Carolina, under Senf, and the Conewago Canal in Pennsylvania, under Brindley, had locks constructed of bricks—unusual at a time when most canal locks were being constructed of wood or stone. See Kapsch, "The Conewago Canal."

31. Wade, "Mount Dearborn," 210.

32. Benjamin Henry Latrobe (1764–1820), architect and engineer, was born in Fulneck, Yorkshire, England, and died September 3, 1820, in New Orleans. He studied both architecture and engineering and worked in England under John Smeaton, considered the father of civil engineering. After the death of his wife, Latrobe immigrated to the United States, arriving in Norfolk, Virginia, in 1796. In the United States he worked on many notable architectural and engineering projects including the Dismal Swamp Canal, the Philadelphia Water Works, improvement of the navigation of

the Susquehanna River, the Washington City Canal, the Chesapeake and Delaware Canal, the U.S. Capitol, and many other projects. See American Society of Civil Engineers, *A Biographical Dictionary of American Civil Engineers,* 1:77–78; and Hamlin, *Benjamin Henry Latrobe.*

33. Robert Morris, president of the Schuylkill & Susquehanna Navigation Company of Pennsylvania, brought English canal engineer William Weston (c. 1752–1833) to Philadelphia in November 1792 to supervise canal design and construction operations. Weston was the son of Samuel Weston, an engineer who had worked for James Brindley, the famous English canal engineer. Before he left for America, Weston had been involved in a secondary role in the construction of the Oxford Canal and as the principal engineer and managing contractor for the design and construction of the Trent Bridge at Gainsborough (1787–91). While in America he consulted on the Potomac Canal and the Middlesex Canal, as well as for the Western Inland Lock Navigation Company and other engineering projects. See Kirby, "William Weston and His Contribution to Early American Engineering," and Skempton, et al., eds., *A Biographical Dictionary of Civil Engineers in Great Britain and Ireland,* 1:773–74.

34. Letter from Benjamin Latrobe to Thomas Moore, January 20, 1811, in *The Correspondence and Miscellaneous Papers of Benjamin Henry Latrobe,* edited by Van Horne et al., 3:17.

35. Letter from John Christian Senf to General Thomas Sumter, Draper mss, Sumter Papers, vol. 1, no. 75, Wisconsin Historical Society, as quoted by Phillips, *A History of Transportation in the Eastern Cotton Belt to 1860,* 37–38.

36. The manuscript version of Senf's November 11, 1800, report, in Senf's handwriting, is at the South Carolina Historical Society. Printed versions appear in U.S. Senate, Ex. Doc. no. 60, *Letter from the Secretary of War,* 11–25, and in Webber, ed., "Col. Senf's Account of the Santee Canal." (The report is hereafter referred to as the Senf report. Pagination cited is from the original handwritten manuscript. There are minor differences between the original and the printed versions. Quotations in this book come from the original manuscript.)

37. Senf's report on the Santee Canal, dated November 11, 1800, includes earlier documents: an extract of a letter from Senf to the president of the Santee Canal Company, dated July 7, 1800; a letter from Senf [to the directors?], dated February 18, 1800; reports from Senf [to the directors?], dated April 2, 1787, and January 10, 1796; and letters from Senf to the president of the company, dated July 24, 1792, and January 18, 1799. The plans accompanying this report, "the general plan of the canal and thirty-eight particular plans on a large scale," have not been found.

38. Reproduced in U.S. Senate, Ex. Doc. no. 60, *Letter from the Secretary of War* and in Porcher, *History of the Santee Canal.*

39. The copy was made by J. P. Gaillard, n.d. The original has not been located.

40. Collections of Drayton Hall, National Trust for Historic Preservation, Charleston, South Carolina.

41. A. S. Salley Jr., appendix in Porcher, *History of the Santee Canal.*

42. "The Santee Canal," in *Report of the Superintendent of Public Works for the Year 1823,* reprinted in Kohn, *Internal Improvement,* 257–69; and [Blanding], "Report of the

Inspection of the Santee Canal, May 7, 1823," South Carolina Department of Archives and History, S208004, General Correspondence of the Superintendent and Board, 1820–48 [hereafter cited as memorandum of visit (1823)]. Kohn's *Internal Improvement* is a valuable compilation of official State of South Carolina documents pertaining to internal improvements undertaken in 1817–28. This book includes the annual reports of the various organizations that were responsible for South Carolina transportation improvements, including the civil and military engineer of South Carolina (1818–19), the South Carolina Board of Public Works (1820–22), and the South Carolina superintendent of public works (1823–28). The book also includes reprints of other pertinent documents and maps.

43. U.S. Senate, Ex. Doc. no. 60, *Letter from the Secretary of War.*

44. Simmons and Newell, *The Santee Canal Sanctuary,* part 1, and Newell and Simmons, *The Santee Canal Sanctuary,* part 2.

45. Bennett and Richardson, "History of the Santee Canal: 1785–1939." I am indebted to Ms. Mary S. Bell of Old Santee Canal Park for providing me a copy of this report.

46. Porcher, *History of the Santee Canal.*

47. Senf report, enclosure, "The Report of the 2d of April 1787," 34.

48. Senf wrote, "Many of the Informations concerning Biggin Swamp were given to me by the Inhabitants at and near it" (Senf report, 33).

49. U.S. Senate, Ex. Doc. no. 60, *Letter from the Secretary of War,* 2. White Oak Landing was also known as White Oak Bluff. Senf report, 2.

50. Senf wrote: "The width of the Canal on the Surface to be 32 Feet, its depth 4 feet, and to contain three feet water; a drawpath on one side ten feet wide, and a Dam of six feet on the other side with counter Ditches on both sides of the Dams (if required) to receive the strange waters and conduct them to their Drains" (Senf report, enclosure, "The Report of the 2d of April 1787," 35).

51. Senf report, 1. Senf described the canal in 1800: "The Canal is twenty two Miles long—thirty five feet wide, five feet and a half deep, four feet water and a Drawpath, ten feet wide; on each side—And has two double—and Eight single—Locks—two large and two smaller Basons with Warehouses, Eight Aqueducts of different Size for small Streams of water to pass under the Bed of the Canal" (ibid.).

52. Senf report, enclosure, "The Report of the 2d of April 1787," 35. When compared to other American canals, the Santee Canal had diminutive dimensions, but it was not necessarily the smallest. The Union Canal in Pennsylvania, for example, had some smaller dimensions. It was thirty-six feet wide at the top, twenty-four feet wide at the bottom of the canal prism, and only four feet deep with locks eight-and-a-half feet wide and seventy-five feet between gates. Like the Santee Canal, the Union Canal was a summit-level canal, and the diminutive dimensions were intended to save water at the summit.

53. Senf recommended, "The size of the Boats for the Canal, should have no more than fifty feet length, and to be nine feet wide" (Senf report, 35). Typical boats using the Santee Canal were somewhat larger, fifty-six feet long, and could carry 80 to 120

bales of cotton. See *Report of the Superintendent of Public Works for the Year 1823*, reprinted in Kohn, *Internal Improvement*, 267. See table 2.2.

54. Senf report, enclosure, "The Report of the 2d of April 1787," 35. The dollar equivalent was calculated using the conversion factor of $4.29 to one pound and is based on 297 pounds, 9 shillings, and 9 pence of South Carolina money being the equivalent of $1,275, as stated in Santee Canal Company, "Report of the Board of Directors" (1805), 5. In 1881 the U.S. Army used a much higher conversion rate, one pound South Carolina money equivalent to five dollars, and calculated that the original cost estimate was $278,000. See U.S. Senate, Ex. Doc. no. 60, *Letter from the Secretary of War*, 5. The actual cost was more than double this estimate.

55. In 1792 William Loughton Smith wrote that Senf was in Philadelphia selling company shares; Rogers, ed., "The Letters of William Loughton Smith to Edward Rutledge (1789–94)," 242, as quoted by Wade, "Mount Dearborn," 210n8.

56. "Notice. Matters of Importance," *City Gazette*, April 1, 1791, 4.

57. "Notice. As a Most Flattering Prospect Is Opening," *City Gazette*, October 13, 1791, 4.

58. "The following rules were agreed to," *City Gazette*, February 1, 1792, 2.

59. "Santee Canal Company," *City Gazette*, January 1, 1800, 3. The newspaper announcement read in part: "At a Meeting of the Directors, the 29th of June 1799, 'Resolved, That on the third Monday in January next, an additional sum of Five Pounds on each share be required from the stockholders, over and above the sum then to be paid agreeable to the tenth rule.'"

60. "Santee Canal," *City Gazette*, June 26, 1797, 3.

61. Senf report, 3–4.

62. Ibid., 2.

63. Memorandum of visit (1823). The memorandum stated, "In 1818 & 1819, the Santee became so low that the sill of the lock would not be passed." The structure built was described as "a wooden form . . . constructed below with a drop which enabled boats to pass." This was probably a wooden crib dam with a flash lock to allow downstream-bound boats to pass.

64. Senf report, 4. Senf provided several different dimensions for the overall length of the lock chambers, such as sixty feet versus sixty-six feet: "The Lock No. 2 is ten feet wide and has ten feet rise—66 feet between the lower and upper gates" (Senf report, 4). Most likely the locks were sixty feet long: "Of course, the Gates of the Locks will be only 60 feet apart and the width of the Locks between 9½ and 10 feet" (Senf report, enclosure, "The Report of the 2d of April 1787," 35).

65. There is some speculation that the bricks were sun dried, although this seems unlikely. See Simmons and Newell, *The Santee Canal Sanctuary*, part 1, 18. More likely they were burned on the spot in ricks, mounds of unburned bricks with a space in the interior for a fire for burning.

66. For a description of the tide lock, see ibid., 46–52.

67. The quarry has been reported to have been in St. Stephen's Parish (now incorporated into present-day Berkeley County) near Pineville. See ibid., 18.

68. Ibid.

69. Despite Senf's use of counterforts there is some evidence that the facing lock walls tipped toward each other. Maxcy Gregg, in his description of an 1839 hunting trip, told how one of their boats had to have its gunwales "hewed" to fit through the lock. This suggests that the lock walls had moved or had tilted toward one another over time. This also occurred on other canals, such as the Chesapeake and Ohio Canal. On that canal, because the lock walls were made of stone, the lock walls and not the boats were sometimes "shaved" to allow passage of boats. See Gregg, "Hunting Journal," 1:6–29. I am indebted to Alexander Moore, University of South Carolina Press, for bringing this manuscript at the South Caroliniana Library, University of South Carolina, to my attention.

70. "The locks, with the exception of the tide-lock, are in a fair condition, though all need some repairs. The tide-lock has been entirely destroyed and all the gates of all the locks are decayed and valueless" (U.S. Senate, Ex. Doc. no. 60, *Letter from the Secretary of War,* 11).

71. Senf reported, "Over every Lock of the Canal is a Bridge sixteen feet wide, besides seven public and private Communication Bridges, without a single Draw—or Turning Bridge" (Senf report, 2).

72. Senf report, 4. Elsewhere Senf indicated that the canal here was forty-five feet wide instead of forty feet. In his report of 1796, he stated that the canal at this location, "is in general 45 feet [wide] on the Surface, 30 feet at the Bottom and on an average eight feet deep, besides the Banks, who are raised 2 feet higher" (Senf report, enclosure, "My report of January 10, 1796," 38).

73. Senf report, 4.

74. See Senf report, enclosure, "The Report of the 2d of April 1787," 35.

75. Senf report, 5.

76. Ibid.

77. The thirty-five foot width seems to contradict other Senf statements that the canal prism was from thirty-two to thirty-five feet. It is possible, as occurred on other canals, that the canal prism varied in dimensions based on the soil conditions through which it passed.

78. Memorandum of visit (1823).

79. An east-west road paralleling the Santee River, the Eutaws Road, was located north of the St. Stephen's Basin of the Santee Canal.

80. Senf report, 6.

81. Senf described the Great Reservoir: "On the left Bank [*sic:* right bank] lays the first or great Reservoir, in its Bank are two strong Floodgates, with an Iron Grate before, arched over and enclosed in a small strong Brickhouse—through which the Canal may be supplied with Water—the Reservoir is here, four hundred yards wide—and the Bank from one Edge of the Swamp to the other is seventy feet thick at the Bottom, well secured with Clay in the Center Ditches and is forty [feet] wide on the Top—and Level with the highland of the sides of the Swamp—The Surface of the canal is two feet lower than the Bottom of this Reservoir" (ibid., 7–8).

82. U.S. Senate, Ex. Doc. no. 60, *Letter from the Secretary of War,* 3.

83. Senf report, 7.

84. "The Santee Canal," *Report of the Superintendent of Public Works for the Year 1823*, reprinted in Kohn, *Internal Improvement*, 257. Chains, also known as Gunter's Chains, were an early measurement system devised by Edmund Gunter. One chain equals sixty-six feet. Each chain has one hundred links, so each link equals .66 feet. Adding up Senf's dimensions of this level provides a distance of 383 chains 25 links or 25,294.5 feet (4.79 miles). For a discussion of the invention and logic of the Gunter Chain, see Linklater, *Measuring America*, 13–16.

85. Senf report, 10.

86. Ibid.

87. Ibid., 11–12. Blanding places the length of this level at "about 1 3–4 mile long" ("The Santee Canal," *Report of the Superintendent of Public Works for the Year 1823*, reprinted in Kohn, *Internal Improvement*, 258).

88. Senf report, 12.

89. Ibid.

90. Ibid.

91. "The Santee Canal," *Report of the Superintendent of Public Works for the Year 1823*, reprinted in Kohn, *Internal Improvement*, 258.

92. Memorandum of visit (1823).

93. Senf report, 14.

94. Ibid.

95. Ibid., 13. Blanding reported the length of this level as somewhat longer, 20 chains (1,320 feet or 0.25 miles). Memorandum of visit (1823).

96. Senf report, 15.

97. Memorandum of visit (1823).

98. Senf report, 17.

99. Memorandum of visit (1823).

100. Senf report, 19.

101. Ibid.

102. Ibid., 20.

103. Ibid.

104. Memorandum of visit (1823).

105. Senf report, 22.

106. Simmons and Newell, *The Santee Canal Sanctuary*, part 1, 48.

107. In 1797 Senf advertised: "Wanted, Forty Labourers, To work on the Santee Canal. The Wages are 20l. per year; one half Men, and one half Women will be received" (*City Gazette*, March 28, 1797). The number "20l" probably refers to twenty pounds in South Carolina currency, or about $84.60 per year.

108. Kapsch, *The Potomac Canal;* Kapsch, "The Potomac Canal: A Construction History."

109. Senf report, enclosure, "The Report of the 2d of April, 1787," 37. The exchange rate is calculated at $4.29 being the equivalent of one pound.

110. Senf report, enclosure, "My report of January 10, 1796," 42–43.

111. Depesseville subsequently relocated to Charleston, where he offered his services for drawing plantations and houses: "Mr. Depesseville. Who for many months has been employed as assistant engineer at the Santee Canal, under col. Senf's direction,

gives notice to the Public in general, that on account of not enjoying his health in that part of the country, he offers now his services for drawing out Plans of Plantations, in colors; also Plans of Houses in profile and elevation, in an exact manner, which will be satisfactory and useful to those who chuse to insure their property" (*City Gazette,* April 8, 1796). Not much else is known of Depesseville.

112. Senf used sixteen- to twenty-foot-long pine planks, two inches thick, for this purpose. See Senf report, enclosure, "My report of January 10, 1796," 44.

113. *Report of the Superintendent of Public Works for the Year 1823,* reprinted in Kohn, *Internal Improvement,* 263. Senf, on the other hand, estimated that a slave could dig and remove somewhat more earth, 70 cubic feet, or about 2.6 cubic yards, per day. See Senf report, enclosure, "My report of January 10, 1796," 42.

114. Senf recuperated in Salem, North Carolina. A gentleman in Camden, South Carolina, reported to a friend in Charleston, "I am happy to have it in my power to inform you that we have news from Col. Senf, as late as the 4th instant, by which we learn his intention of setting off from Salem, North Carolina, where sickness has been the cause of his long stay, in two or three days after the date of his letter, for the Canal. This news, I am satisfied, will be very acceptable to you and your friends who own shares in this undertaking, which will receive new vigor after the return of the Colonel to the Canal through his presence" (*City Gazette,* January 21, 1797).

115. Senf report, enclosure, "The Report of the 2d of April, 1787," 36.

116. Ibid.

117. Typhoid fever can also be transmitted by means other than impure water, including by human carriers, contaminated milk, direct contact, and flies. Impure water was, however, probably the leading transmitter for workers of the Santee Canal Company.

118. The Center for Disease Control (CDC) Web site states: "Typhoid fever is a life-threatening illness cause by the bacterium *Salmonella Typhi.* . . . You can get typhoid fever if you eat food or drink beverages that have been handled by a person who is shedding *Salmonella Typhi* or if sewage contaminated with *Salmonella Typhi.* Typhi bacteria gets into the water you use for drinking or washing food. . . . Typhoid fever is an acute, life-threatening febrile disease. . . . The hallmark of typhoid fever is persistent, high fevers. Other common symptoms and signs include headache, malaise, anorexia, splenomegaly, and relative bradycardia." See http://www.cdc.gov (accessed May 6, 2009).

119. The CDC currently reports four hundred cases of typhoid a year in the United States, 70 percent resulting from travelers who contract the disease in developing countries.

120. Puddling was a technique of installing an impervious layer of tamped clay to seal the canal prism and limit water leakage.

121. John Drayton, *A View of South-Carolina,* 32–33.

122. Most of these officers and directors had a long involvement with the company. The January 18, 1797, issue of the *City Gazette* reported that the stockholders had elected David Ramsay as president; Edward Rutledge, Nathaniel Russell, Thomas

Jones, Edward Darrell, John F. Grimké, John S. Cripps, John Rutledge Jr., J. F. Chion, Thomas Morris, David Campbell, Henry Laurens, and A. Gilchrist as directors; and John M. Davis as secretary.

123. The public report of the February 28, 1797, board of directors meeting read:

> Yesterday the Stockholders in the Santee Canal Company had a meeting, at which col. Senf, the director of the work attended, and gave very minute and satisfactory information and statements respecting the same. From his communications it appeared that the unfinished part of Canal lies between Stephen Mazyck's, and Flint's, a distance of between five and six miles, the whole of which, except one mile, is a dry soil, easily cut through; and that all the locks, aqueducts, and the digging work, may be finished by next January, with the addition of sixty laborers and seventeen tradesmen to the number at present actually employed on the Canal, so that boats may pass from Santee to Cooper river.
>
> The additional force requested by col. Senf was unanimously voted by the Company. Well founded hopes may therefore by indulged, that the crop of the present year will be transported from one river to the other through the intermediate country, and that heavy laden boats gliding through the forest will soon occupy the identical spots on which the growing pine trees lately raised thier lofty heads.
>
> A boat, constructed by Mr. Chesnut of Camden, agreeably to a plan directed by col. Senf, for passing through the Canal from one river to the other without unloading, is now at the north end of the Canal. This is fifty-one feet long, and nine feet wide, and will carry 900 bushels of corn, or two hundred barrels of flour, or thirty-five hogsheads of tobacco. Boats of this size can be built and navigated at half the expence, and will perform two trips to one of those that go round by the mouth of the Santee.
>
> The canal boats, drawing only two and an half, or at most three feet water, can go up and down the Santee in the driest times, and without being much affected by the wind; whereas the boats now in use, can only be used when there is a swell in the river, and are often detained for several days, by bad winds, from passing Bull's Bay. At present, and for some time past, they have not been able to bring down a sufficient supply of corn for the Charleston market, from the want of water to float them. When the canal is completed, these inconveniences will be done away, and the valuable productions of the back country will be poured into Charleston with such regularity and uniformity, that the arrival at, or the return of a boat from Camden or Granby, may be calculated nearly to a day, with precision and certainty. (*City Gazette*, March 1, 1797, 3)

Despite the optimism of this report, it took another three years to complete the canal.

124. See Kapsch, *The Potomac Canal*, and Kapsch, "The Potomac Canal: A Construction History."

125. See Kapsch, "The Conewago Canal."

126. They were paid as follows:

Superintendent	$1,200/year plus house
Lock Keepers	$300/year plus house (except keeper for lock 3, who was paid $400/year)
Toll Receivers	$500/year plus house
Boom Keeper	$300/year plus house

See *City Gazette,* April 30, 1800, 3.

127. "Santee Canal," ibid., April 12, 1800, 1.

128. Ibid.

129. In 1823 a slave named Stephon was hanged near the Santee Canal, not for destroying canal property but for being part of a band of runaway slaves turned bandits. The October 14, 1823, issue of the *Southern Patriot,* quoted a letter sent from Pineville the previous day reporting on the actions taken against the bandits:

> A party went out on Tuesday last, and routed the remainder of Joe's gang. They fired on two of the men and wounded one. One of the Negroes wounded Evance Cordes in the shoulder and one finger, with duck shot. The wound is very slight and he is entirely recovered. On Wednesday, some of the same party went out, and came up with two men and three women. One of the men fired on John Ravenel, but without effect, and made his escape. They took one fellow and the wenches, and we have them now in confinement in this place. One of the women having a man's cloak on was fired upon, and I am sorry to say was wounded—her chance of recovery is doubtful. On Tuesday when the party were routed, they left a child about three years old in the camp, and a chance shot struck it in the back of the head, and it has since died. The fellow taken and the woman wounded, (who is the wife of Joe) belongs to Dr. Roual; a Girl, the mistress of Joe, belongs to Mr. Ballard; and the other woman, the wife of Anderson who was killed, belongs to Mr. Bates. Two of the most desperate and one woman, are not yet taken.
>
> A Court of Magistrates was sitting for the trial of those taken, and others suspected. An Acting Committee of the Association are very active in gaining information.

A week later the October 21 issue of the *Southern Patriot,* published an extract from a letter sent from the Santee Canal the previous day reporting the outcome: "I am happy to inform you that all of the Company's Negroes are discharged; and declared innocent. Two of them only received a moderate correction, which however was not for being found guilty of having any connexion with those runaway negroes, but merely for their inconsistency in their evidence. One of Joe's gang named Stephon, belonging to Dr. Roual, was hung close by the Canal Bridge, on Friday last; and I hope this unpleasant affair may be finally suppressed now."

130. *Times* (Charleston, S.C.), May 28, 1800, quoted in Porcher, *History of the Santee Canal.* Porcher gives the date of this article as 1801 but this is most likely a typographical error.

131. Senf based his recommended tolls on a typical boat weighing 40,000 pounds at a toll rate of three pence a hundred pounds: "To establish the Toll of the Boats, which

pass the Canal at the Value of the Cargo, would not be [in] the Company's Interest; and [would] be connected with a number of Difficulties, as the Boats mentioned before will carry above forty Thousand weight; only these forty thousand weight, at three pence per hundred; each Boat would pay five pounds in passing the Canal, without inquiring what its Cargo [was]. The Owners of Boats will always take care to have their Boats loaden [fully loaded], when understood that they are to pay the same Toll when not full loaden" (Senf report, enclosure, Letter from Senf to the president of the Santee Canal Company, January 18, 1799, 50).

132. *Report of the Superintendent of Public Works for the Year 1823*, reprinted in Kohn, *Internal Improvement*, 267, and Santee Canal Company, "Report of Board of Directors" (1805). In his 1808 report on public roads and canals, Secretary of the Treasury Albert Gallatin also reports that the Santee Canal cost $650,667, "including sixty negroes and some tracts of land belonging to the company." [Albert Gallatin], *Report of the Secretary of the Treasury on the Subject of Public Roads and Canals*, 25. Also included in U.S. Congress, *Report of the Committee on Roads and Canals.*

133. Porcher, *History of the Santee Canal.*

134. Ibid. The superintendent of public works made a similar calculation: "The canal, from its commencement in May, 1793, to its completion in 1800, cost $650,667.14. Dividends were made in 2 years only—In 1814, $12,359, and in 1815, $13,086; in all, $25,445, or less than the legal interest on the capital expended, accruing, since the completion of the canal, by more than $966,000—so that the whole expense of the stockholders, including principal and interest exceeds $1,615,000" ("The Santee Canal," *Report of the Superintendent of Public Works for the Year 1823*, reprinted in Kohn, *Internal Improvement*, 267).

135. The official cost stated for the one-mile long Conewago Canal in Pennsylvania was $105,000. The actual cost of this canal was much larger as many of the workmen were never paid for their work. See Kapsch, "The Conewago Canal."

136. Senf report, enclosure, "A letter to the President of the Company, July 24, 1792," 48–49.

137. Senf had recommended that ten horses be stationed at both Santee Basin and at Biggin Basin with twenty African Americans to tow the boats through the canal. Because the boats had large crews, who were able to pull their boats through the canal, these plans were abandoned. See Porcher, *History of the Santee Canal.*

138. "The Santee Canal," *Report of the Superintendent of Public Works for the Year 1823*, reprinted in Kohn, *Internal Improvement*, 267.

139. [Albert Gallatin], *Report of the Secretary of the Treasury on the Subject of Public Roads and Canals*, 25.

140. The superintendent of public works reported that annual expenses of the Santee Canal Company were $7,000 per year and annual income varied from a low of $11,031 in 1822 to a high of $21,249 in 1814. See "The Santee Canal," *Report of the Superintendent of Public Works for the Year 1823*, reprinted in Kohn, *Internal Improvement*, 267.

141. In 1810 Latrobe advised William Loughton Smith: "To supply a canal by pumping is one of the most desperate remedies which the evil of a scanty supply of water

can require. Do not *treble it* without absolute necessity and an overflowing treasury" (letter from Benjamin Latrobe to William Loughton Smith, February 13, 1810, *The Correspondence and Miscellaneous Papers of Benjamin Henry Latrobe,* 2:844).

142. U.S. Senate, Ex. Doc. no. 60, *Letter from the Secretary of War,* 6.

143. Ibid., 3.

CHAPTER 3. *The Catawba Company, the U.S. Military Establishment, and Early South Carolina Canals and Waterways, 1787–1812*

1. Rocky Mount no longer exists. It was situated immediately south of Great Falls, South Carolina, on the west bank of the Catawba River.

2. Letter from President George Washington to, Secretary of War James McHenry, December 13, 1798, in *The Papers of Alexander Hamilton,* 22:352, cited by Wade, "Mount Dearborn," 209n5.

3. Henry Dearborn (1751–1829), soldier, secretary of war, and diplomat, was born in North Hampton, New Hampshire, and educated at public schools. During the Revolutionary War, he distinguished himself in many engagements, including the Battle of Bunker Hill, the storming of Quebec, the Battle of Saratoga, the Battle of Monmouth, and the siege of Yorktown. He began military service as a captain and was promoted to colonel. After the war he was elected brigadier general of the militia (1787) and later made major general (1789). He was elected to Congress, where he served from 1793 to 1797. In 1801 President Thomas Jefferson appointed Dearborn secretary of war, and he served in that position until 1809. Subsequently he held many other government positions, including senior major general of the U.S. Army and minister to Portugal. See *Biographical Directory of the United States Congress,* http://bioguide.congress.gov (accessed April 30, 2009).

4. Thomas Sumter (1732–1833), soldier, legislator, and diplomat, was born in Hanover County, Virginia. About 1760 he moved to South Carolina. During the Revolutionary War he was appointed a lieutenant colonel and was promoted to brigadier general of militia in 1780. He distinguished himself in many military actions throughout South Carolina. After the war he served in several elected positions, including U.S. congressman (1789–93 and 1797–1801) and as U.S. senator (1801–10). Subsequently he served as minister to Brazil. Fort Sumter, in Charleston harbor, was named in his honor. See *National Cyclopedia of American Biography,* 79–80; and *Biographical Directory of the United States Congress,* http://bioguide.congress.gov (accessed April 30, 2009).

5. Letter from Secretary of War Henry Dearborn to General Thomas Sumter, May 26, 1802, National Archives, Record Group 107, Miscellaneous Letters Sent, Office of the Secretary of War (microfilm M370/1).

6. Letter from Dearborn to Sumter, June 5, 1802, ibid.

7. Dearborn wrote to Sumter: "The quantity of land must in some measure depend on the terms on which it can be purchased, twenty or thirty acres well situated would be sufficient for all the necessary purposes, but if a tract of [indistinct] land adjoining could be had on low terms it may not be improper to purchase fifty acres, but as our means are small at present we must use them with economy" (ibid.).

8. Dearborn wrote: "Will you Sir undertake to select the Scite with the aid of Colonel Senf (who I have requested to assist you) and make the purchase of the land including the water privilege necessary for the establishment" (ibid.). On the same day Dearborn wrote to Senf: "I have this day written to General Sumter requesting him to undertake in conjunction with you on the part of the U. States to purchase a Scite for a Magazine and Armoury at Rocky Mound on the Catawba River. I have therefore to request if your engagements will permit, that you will undertake this business as soon as convenient for both of you to attend" (letter from Dearborn to John Christian Senf, June 5, 1802, ibid.). At the time Senf was living in the nearby hamlet of Rocky Mount.

9. Letter from Dearborn to Sumter, June 5, 1802, ibid.

10. The other two were established earlier: Springfield, Massachusetts, in 1777; Harper's Ferry, Virginia (now West Virginia), in 1799. See Wade, "Mount Dearborn."

11. Letter from Dearborn to Sumter, August 9, 1802. National Archives, Record Group 107, Miscellaneous Letters Sent, Office of the Secretary of War (microfilm M370/1).

12. Wade, "Mount Dearborn," 213.

13. Ibid., 213–14.

14. Letter from Dearborn to Senf, November 12, 1802. National Archives, Record Group 107, Miscellaneous Letters Sent, Office of the Secretary of War (microfilm M370/1).

15. Ibid.

16. Letter from Dearborn to Senf, February 23, 1803, ibid.

17. Dearborn explained his understanding of the agreement between the army and the canal company to company president John F. Grimké: "From the verbal and written information received at different times from Col. Senff, I considered the completion of the Canal and a supply of water from it for our works as a matter to be relied on: and the expense of a dam was authorized accordingly" (letter from Dearborn to Grimké, August 5, 1807, ibid.).

18. Kapsch, "George Washington, the Potomac Canal and the Beginning of American Civil Engineering History: Problems and Solutions"; Kapsch, "The Potomac Canal: A Construction History"; Kapsch, *The Potomac Canal,* 116–17.

19. John Faucheraud Grimké (1752–1819), soldier, legislator, and jurist, was born in Charleston, South Carolina, and educated at Westminster School and Trinity College, Cambridge University, receiving his bachelor of arts degree in 1774. He returned to South Carolina in 1775. During the Revolutionary War, Grimké entered the regiment of artillery as a lieutenant, and was later promoted to captain, major, and then lieutenant colonel. He participated in the Battle of Stono, the siege of Savannah, and the fall of Charleston. Later in the war, he joined General Nathanael Greene's Continental Line. In 1784 Grimké married Mary Smith, and they subsequently had fourteen children. Grimké served in the South Carolina House of Representatives for five years. In 1783 he was appointed an associate judge, and in 1799 he was appointed senior associate judge, a position he held until his death. He was the author of several legal studies. He was active in the early development of internal improvements in

South Carolina, becoming president of the Catawba Company in 1802. See Bailey and Cooper, *Biographical Directory of the South Carolina House of Representatives,* 3:290–92; and *Dictionary of American Biography,* 4:633–34.

20. Letter from Senf to Grimké, March 30, 1803, copied into the letter from Grimké to Capt. Alexander Macomb, July 3, 1807, and forwarded by Macomb to Henry Dearborn, July 22, 1807, National Archives, Record Group 107, Letters Received by the Secretary of War (microfilm M221/10).

21. In 1807 at the height of the dispute between the army and the Catawba Company, Macomb wrote to John F. Grimké: "I am also commanded to remind the Company of their engagement to have the passage for the water opened by the 1st day of October 1803, if practicable, and as it is not yet done, to enquire when they contemplate effecting it" (letter from Macomb to Grimké, June 9, 1807, ibid.).

22. Letter from Dearborn to Grimké, ibid., Miscellaneous Letters Sent, Office of the Secretary of War (microfilm M370/1).

23. Letter from Dearborn to Senf, May 14, 1803, ibid.

24. Dearborn had provided Eli Whitney with a detailed description of the desired buildings:

Memorandum

Buildings for the establishment at Rocky Mount in South Carolina.

The principal Building for the Armoury to be 80 feet by 30, two stories high; lower story 12 feet, upper Story 10 feet.

Building for a Trip Hammer and for forging Barrels etc. 50 feet by 30—one story of 12 feet.

Building for Stocking, finishing & polishing. 64 feet by 26—two stories high, of 10 foot each.

Barracks for fifty Workmen and a guard of twenty Men

One Story and an half high, the latter for lodging rooms; say 100 feet in length, and 18 feet in width; and two of 55 feet each in length, and 18 feet in width—One story 10 feet; and the half story 4 feet.

A House for the Superintendant & Paymaster & Store-Keeper a 38 feet by [indistinct] two stories high, each Story 9 feet.

An entry and Strain-Way in the Center rear rooms for Kitchen

The whole of the Walls of the several buildings to be stone or brick. The Carpenter's work to be done plain & strong. The Barracks to have a Piazza in front 8 feet in width.

A suitable site should be designated for a Building 100 feet by 32 for the reception of Muskets. And also a site for a Powder Magazine. (letter from Dearborn to Whitney, November 7, 1803, ibid.)

25. William Loughton Smith (1758–1812), legislator, diplomat, and president of the Santee Canal Company, was born in Charleston, South Carolina, and educated in England. He was a South Carolina Federalist and a strong promoter of internal

improvement in South Carolina and elsewhere. He invested heavily in both the Catawba Company and the Santee Canal Company. He served in the South Carolina House of Representatives (1784–88, and 1808) and the U.S. House of Representatives (1789–97) and was minister to Portugal in 1797–1801. Smith later became president of the Santee Canal Company. See *The Correspondence and Miscellaneous Papers of Benjamin Henry Latrobe,* 1:393n; *Biographical Directory of the United States Congress,* http://bioguide.congress.gov (accessed April 30, 2009); and Rogers, *Evolution of a Federalist.*

26. John Vaughan (1756–1841), Philadelphia merchant, was a member of the American Philosophical Society and virtually every other learned and benevolent society in Philadelphia. See *The Correspondence and Miscellaneous Papers of Benjamin Henry Latrobe,* 1:229n.

27. Letter from Latrobe to Vaughan, March 18, 1804, *The Correspondence and Miscellaneous Papers of Benjamin Henry Latrobe,* 1:455–56.

28. Frederick Graff (1774–1847), engineer, was the son of Jacob Graff, a bricklayer and was apprenticed as a carpenter. In 1794 Latrobe hired Frederick Graff as an assistant engineer for erecting the first Philadelphia Waterworks. He became Latrobe's draftsman and assistant at the waterworks (1797) and for the Bank of Pennsylvania. Most of his career was spent at the Philadelphia Waterworks, building the first steam-powered waterworks in the United States at Centre Square (1801) and planning and constructing Fairmount as the site for the new water reservoir (1810). See *The Correspondence and Miscellaneous Papers of Benjamin Henry Latrobe,* 1:456n; and American Society of Civil Engineers, *A Biographical Dictionary of American Civil Engineers,* 1:51.

29. Letter from Latrobe to Vaughan, March 18, 1804, *The Correspondence and Miscellaneous Papers of Benjamin Henry Latrobe,* 1:455.

30. Ibid.

31. Letter from Latrobe to Vaughan, March 21, 1804, and to William Loughton Smith, March 21, 1804, ibid., 1:457–59.

32. We do not have Grimké's letter to Dearborn, but we do have Dearborn's response: "I am happy in hearing that you are progressing with the Canal near Rocky mount and that what was intended to have been accomplished last year will probably be executed this" (letter from Dearborn to Grimké, July 12, 1804, National Archives, Record Group 107, Miscellaneous Letters Sent, Office of the Secretary of War [microfilm M370/2].

33. Graff returned to Philadelphia to superintend the design and construction of the Fairmont Waterworks for the City of Philadelphia.

34. *The Correspondence and Miscellaneous Papers of Benjamin Henry Latrobe,* 1:459–60n.

35. Letter from Dearborn to Lieutenant Colonel Francis Mentges, March 7, 1805, National Archives, Record Group 107, Miscellaneous Letters Sent, Office of the Secretary of War (microfilm M370/2); letter from Dearborn to Senf, March 9, 1805, ibid.

36. Wade, "Mount Dearborn," 221.

37. Ibid.

38. Alexander Macomb (1782–1841), soldier and general-in-chief of the U.S. Army, was born in Detroit, then a frontier garrison town, and soon moved with his family to New York. He was educated at the Newark Academy in New Jersey. In 1799 he received a commission as a cornet of light dragoons. He married his cousin, Catharine Macomb, in 1803. In 1805 he was appointed captain in the Army Corps of Engineers. In 1806 he was assigned as superintendent to the U.S. military establishment on the Catawba River, then being constructed. The following year he was reassigned to plan the fortifications of the coast of Georgia, North Carolina, and South Carolina, a position he held until reassigned in 1811. He advanced quickly in the army during the War of 1812 and by 1814 was appointed brigadier general. In 1821 he was placed in charge of the army's Engineer Bureau, and in 1828 he was made general-in-chief of the U.S. Army. See Jenkins, *Daring Deeds of American Generals*, 295–322.

39. Dearborn's termination letter to Senf reads in part: "It has been deemed expedient to send Capt. Macomb, of the Corps of Engineers, to take upon himself the direction of the Public Work at Rocky Mount. You will therefore please to deliver over to him all public property in your possession, together with such instructions and drawings, as you have received relative to the proposed Establishment;—and you will, with as little delay as practicable, transmit your accounts with proper vouchers for a final settlement. On receipt of this, you will consider your Agency, under the United States, for superintending the public works at Rocky Mount, as ceasing and will render your account for service accordingly" (letter from Dearborn to Senf, April 18, 1806, National Archives, Record Group 107, Miscellaneous Letters Sent, Office of the Secretary of War [microfilm M370/2]).

40. Wade, "Mount Dearborn," 222. Senf's obituary read: "Died, at his Seat in Rocky Mount, South Carolina, on the 24th ult. of a lingering illness, in the 53rd year of his age, Colonel Christian Senf, Chief Engineer of the State of South Carolina. He was an Officer of merit and information, and had served with great applause in the Southern States, as an Engineer, during our Revolutionary contest" (*Charleston Courier*, September 3, 1806, quoted in appendix to Porcher, *History of the Santee Canal*).

41. Dearborn's instructions on how to build a magazine at Rocky Mount were quite detailed and took into account the danger of explosions:

> Directions relating to a Magazine to be erected at Rocky Mount, S.C.
> First—erect a circular frame building, eighteen feet in diameter, inside; and twelve feet high exclusive of the door or roof. The scatling to be six inches square, and boarded high on both and the space between the inside & outside boards to be filled in with charcoal beat [with] a mall. The roof is to form a cone, and also to be boarded on both sides. The floor is to be laid on scatling eight inches square, well supported, and, the spaces between the scatling to be filled with fine charcoal, sitting on as under floor. The under side of the scatling for the floor is to be eighteen inches above the earth.—There should be one glass window on each side of the door, about two and a half feet square, to be on hinges. The Door should not be more than three feet in width & six & a half feet high. The Hinges of the door and windows should be of copper or composition.

Round this frame building should be erected a circular brick building, terminating in a cone of brick work, capped with a stone. The Brick for this case should be made for the purpose, with a bevel at one end to correspond with the Angle of the cone. The inside of the Brick work should be two & a half feet from the frame building. The walls should be 21 Inches thick, at the base, & 18 Inches at the commencement of the cone or roof. This inside is to be perpendicular, by which means the walls are the support the dome or roof. The Door should not be more than three feet in width, & six and a half feet in height: the frame should be of Stone, the Hinges of composition, and, the outside of the Door covered with sheet copper. There should be four ventilators through the brick wall, at the height of [indistinct] upon the base, made in a zig zag manner through it, & covered on the outside with sheet copper perforated with holes ½ Inch diameter, as close together as practicable, for the purpose of admitting air. The sheet copper perforated in this manner will prevent birds and other mischievous animals from passing into, or filling up the aperture with nests.—A good electrical conductor should be attached to the building, with its upper extremity four feet above the top of the dome or roof—both extremities should be terminated with [indistinct] copper or bronze. (letter from Dearborn to Macomb, April 30, 1806. National Archives, Record Group 107, Letters Sent, Military Affairs, Office of the Secretary of War [microfilm M6/2]).

42. Dearborn's detailed instructions to Macomb for the other buildings to be built at Rocky Mount read:

You will please to have a brick Barrack erected for Rocky Mount, one story high and thirty six feet in length; forming [indistinct] rooms with a large fire place in each. The walls of the Barrack should be continued three feet above the lower rooms, for the purpose of affording bed-rooms above. There should be one window in front, and one on the back of each room, and a [indistinct] window in the roof of each bed room. [Indistinct] of the bottom a piazza should be erected eight feet in width. You will also erect two Brick Houses in one block two stories high, with two rooms on each floor of each house, one in front & one back. The Houses are to be plain and strong; and the whole block should be forty six feet in length, exclusive of outside walls, and thirty three feet deep, exclusive of the walls. A front door and entry should be made to each House; and also a Kitchen. A Shed is also to be erected, one hundred feet in length, with brick walls, seven feet high on the back side and ends; and with six brick columns in front for supporting the roof which is to have a ridge as usual for other buildings to be boarded & shingled; and painted with oil of Spanish brown, as should be the roofs of the other buildings. (letter from Dearborn to Macomb, May 1, 1806, ibid. (microfilm M6/3).

43. In order to recruit skilled workers, Dearborn wrote to William Linnard:

We are in want of the following workmen to go to a place, called Rocky Mount, in the interior part of South Carolina, where the public are forming an establishment for an Arsenal, magazine & armoury. It is said to be a healthful mountainous place. The workmen will be under the immediate care of Col. Mentses, who set out this day for that place. Viz. One master stone mason, with two or three journeymen. One blacksmith, with a complete set of tools, one carpenter, one judicious practical brick maker, and two or three good faithful men who understand splitting rough stones with powder or with wedges, with suitable tools to split them in the different methods. Generous wages will be given for good men, and the employment will probably continue at least two years. (letter from Dearborn to Linnard, March 12, 1805, ibid.).

This recruitment was successful for on June 6, 1805, Dearborn was informing Senf:

The following tradesmen have been engaged at Philadelphia, to proceed to Charleston & from thence to Rocky Mount, and will sail immediately with the necessary tools etc.—

One Master Mason & Stone Mason	@	$3 pr day	
Three App. Masons	"	1–50/100 pr. day.	
One Master Carpenter	"	3	"
Two Appt. Carprtr.	"	1–50/100	"
One Black Smith	"	1–75/100	"
Three Stone Quarriers	"	1–75/100	"

(letter from Dearborn to Senf, June 10, 1805, ibid., Miscellaneous Letters Sent, Office of the Secretary of War [microfilm M370/2]).

44. Dearborn's instructions to Macomb on handling these escaped slaves read:

"Your letter of the 17th ult. is before me. It will I presume be proper for you to give the earliest notice to the Gov. of Georgia, when any Negroes are delivered to you by the Indians, requesting him to take measures for having them received from you as soon as possible, paying all costs: The negroes can inform you who their respective owners were, which may be useful for you to communicate to the Govr. Should the negroes not be called for in a reasonable time, with evidence of ownership, you may set them to work, taking due care to prevent their escape. You will, at all times, treat them with humanity" (letter from Dearborn to Macomb, August 11, 1806, ibid., Letters Sent, Military Affairs, Office of the Secretary of War [microfilm M6/3]).

45. Postscript to a letter from Macomb to Dearborn, February 27, 1807, ibid., Letters Received, Office of the Secretary of War (microfilm M221/10).

46. Letter from Macomb to Grimké, June 9, 1807, ibid.

47. Letter from Grimké to Macomb, July 6, 1807; forwarded by Macomb to Dearborn, July 22, 1807, ibid.

48. Letter from Senf to Grimké, no date (ca. 1803), quoted in a letter from Grimké to Macomb, July 6, 1807; forwarded by Macomb to Dearborn, July 22, 1807, ibid.

49. Letter from Grimké to Macomb, July 6, 1807; forwarded by Macomb to Dearborn, July 22, 1807, ibid.

50. Ibid.

51. Letter from Macomb to Secretary of War William Eustis, December 8, 1809, ibid. (microfilm M221/27), as quoted by Wade, "Mount Dearborn," 329.

52. Signed by company president John Grimké and dated September 22, 1808, the memorial read:

> *THE PETITION and MEMORIAL of the Subscriber, in behalf of the Company incorporated by law for the opening the Navigation of the Catawba and Wateree Rivers—*
>
> *SHEWETH—*
>
> THAT the present Stockholders of the said Company, convinced of the great advantages which would result from an Inland Navigation, and more particularly from removing the obstructions in the Catawba and Wateree Rivers, whereby the productions of a considerable portion of the States of North and South-Carolina and Tennessee, would be conveyed to Charleston; and believing, from the best information they could obtain, that removing the said obstructions might be accomplished at a moderate expence, and within a reasonable period, did, several years ago, become proprietors of the original shares.
>
> THAT the Company has already expended considerable sums on the said Rivers, and are still annually expending more, with the well-founded hope that, by a continued though small exertion, they may finally accomplish so valuable and desirable a work—but, that from the extent of the River, the number of obstructions, and the small association of persons at present engaged in this important work, there is but little prospect of the obstacles being removed so as to render the navigation of the said Rivers practicable for boats, even of small burthen, for many years to come.
>
> YOUR Petitioners have, therefore, with much pleasure, perceived that the subject of Canals and Roads has attracted the serious attention of the Legislature of the United States, in consequence of the very able and luminous report of the honorable the Secretary of the Treasury, which has given features to that subject, which, if steadily pursued, will prove not only an ornament, but be of infinite importance to these United States.
>
> THAT in the present weak and embarrassed state of the Company, your Petitioners are induced and compelled to resort for pecuniary aid to the Legislature of the United States, which, by a wise and judicious management of its resources, is now fully competent to afford efficient succour and encouragement to works of great public benefit, seldom to be accomplished by individual efforts, as your petitioners have fatally experienced, particularly in this young country. Your honorable body, by authorising the Secretary of the Treasury to purchase, for the use of the United States, all their vacant shares, (which are about a moiety of the whole number) or such part of them as you shall, in your wisdom, direct to be subscribed for the United States, will accelerate and secure the completion of this important navigation, and vest in the United States a highly valuable interest to be derived

from the tolls which would be received from the boats navigating the River, and would thereby become not only profitable to the individual members, but particularly so to the United States, whose public Arsenal and Armory, fixed on the Catawba River between two Falls of the greatest Shoal on the River, cannot prove of as advantageous a result as if the navigation of that River was free and unimpeded.

YOUR Petitioners subjoin a statement, for the information of your honorable body shewing the extent of the said River, the population of the country which it waters, and the advantages which must of course accrue to the United States if the said Rivers should be immediately opened; and relying on your patriotic disposition to promote this valuable work, they flatter themselves their application will be favorably received, and as in duty bound will ever pray. (Library of Congress)

53. The congressional report concluded that "although the committee are fully convinced, that the improvement of the navigation of the said river would be of vast importance and utility to the inhabitants of a considerable portion of the states of North and South Carolina and Tennessee . . . and that the armory and arsenal of the United States, established at Rocky Mount, on the Catawba river, would also be considerably benefitted by the opening of the river: yet, from the present state of our finances, and the critical situation of our country in relation to foreign governments, they are of opinion that it would be unadvisable to apply any public money, at this time, to internal improvements of this kind" (U.S. Congress, *Report of the Committee* [1809]).

54. Weaver, *Internal Improvements in North Carolina,* 65.

55. The complete title was "An Act for Appointing and Impowering Commissioners to Make the Wateree River Navigable, and for Laying out and Making a Road from the Upper Settlements, near the Catawba Nation, on the North-East side of the Said Wateree River, to the Place on Santee River commonly called Beard's Ferry, and from thence to the Road Leading from the Congrees to Charlestown," enacted April 21, 1753, McCord, ed., *The Statutes at Large of South Carolina,* 7:504–6.

56. Ibid., 7:505.

57. Ibid.

58. Ibid.

59. The legislation provided: "And in any case any of the said inhabitants shall neglect or refuse to work in removing the said rafts, or otherwise in making the said Wateree river navigable, or shall neglect or refuse to work in making the said road, or keeping the same in repair, they shall be subject to the same fines and forfeitures as any persons, by Act of the General Assembly for cutting, cleansing and clearing of creeks and water-passages, or making and keeping roads in repair, are made subject and liable unto" (ibid.).

60. Ibid.

61. "An Ordinance to Appoint Commissioners for Clearing Wall's Cut, Edisto, Wateree, Great and Little Peedee Rivers," enacted March 26, 1784, ibid., 7:531–33.

62. "An Act for Opening and Improving the Navigation of Great Peedee, Wateree, Congaree, Broad River, Savannah, Keowe, Tugaloo and Black Rivers, Lynch's, Black, Jeffrie's and Catfish Creeks," enacted February 19, 1791, ibid., 7:561–65.

63. *Report of the Civil and Military Engineer for 1818,* reprinted in Kohn, *Internal Improvement,* A18.

64. Newspaper announcement, Joel R. Poinsett Papers, "Edisto and Ashley River Canal Company: 1787," South Carolina Historical Society.

65. Ibid.

66. "An Act to establish a company for," enacted March 27, 1787. This was the result of a petition to the General Assembly by William Moultrie, John Rutledge, Thomas Bee, James Lynch, Minor Winn, William Butler, John Purvis, William Anderson, John Budd, Leroy Hammond, Ephraim Mitchell, Nicholas Eveleigh, William Saunders, Peter Freneau, David Olyphant, Wadsworth and Turpin, Charles Goodwin, William Moore, Thomas Hall, James Milligan, Stephen Drayton, Christopher Gadsden, Jacob Read, Alexander Gillon, Andrew Pickens, Thomas Brandon, Patrick Colhoun, James Lincoln, and James Mayson. See McCord, ed., *The Statutes at Large of South Carolina,* 7:545.

67. *Report of the Civil and Military Engineer for 1818,* reprinted in Kohn, *Internal Improvement,* A18.

68. Copy of the minutes of the Edisto and Ashley River Canal Company, February 27, 1788. Joel R. Poinsett Papers, "Edisto and Ashley River Canal Company: 1787–1788," South Carolina Historical Society.

69. The company authorized the printing of fifty handbills and the publication of the following advertisement in the *City Gazette:*

> Edisto Company
> The General committee of the incorporated company for Clearing the navigation etc. of the Edisto River—Give Notice—that they are ready to receive proposals from any person or persons who may be inclined to contract for the clearing and making navigable the river Edisto or any part therefore from the Saw Mills of the Late Thomas Furguson Esqr: deceased to the heads of the respective branches of the Same river as far as the navigation can be carried, such proposals to be Sent under Seal Cover Directed to Mr. Adam Gilchrist No. 1 Gillon Street, Charleston, in front of the Exchange or to Lewis Lestarjette Esqr. At Orangeburgh on or before the 18th day of May next. (Ibid.)

70. Letters received, Edisto and Ashley River Canal Company, ibid.

71. Diamonde's proposal read:

> Understanding that its Intended to Cut a Canal between Ashley & Edisto Rivers, for promoting inland Navigation, & as I am, Acquainted with the Situation (as I conceive) thro' which it ought to pass, I would propose to make a Survey & Plan of the same with the elevation of the Edisto above Ashley River. Respecting an estimate of the expense, I have no experimental knowledge of such Business, yet I am Persuaded it is not difficult to Procure Information from Rice Planters & their Managers, experienced in draining rice fields, of the Quantity of Labour performed by a given Number of hands in a given time, for thro' such Ground it must Pass, if carried into execution. Such Information I would endeavour to procure & make the Measurement, Leveling, & Plan for three hundred Dollars,

Including the expence of Assistants, etc. (Letters Received, Edisto and Ashley River Canal Company, John Diamond to the Board, April 3, 1809, Joel R. Poinsett Papers, "Edisto and Ashley River Canal Company: 1801–1809," South Carolina Historical Society).

72. Letters Received, Edisto and Ashley River Canal Company, ibid.

73. *Report of the Civil and Military Engineer for 1818,* reprinted in Kohn, *Internal Improvement,* A18. The second survey was conducted by a Mr. Payne, for Eves and Trotti. See *Report of the Superintendent of Public Works for 1825,* reprinted in Kohn, *Internal Improvement,* 370.

74. [Sampit and St. James Santee Canal Company], *Rules and Regulations;* McCord, ed., *The Statutes at Large of South Carolina,* 7:580–82.

75. *Report of the Superintendent of Public Works for 1823,* reprinted in Kohn, *Internal Improvement,* 271.

76. "An Act Appointing Commissioners to Lay Out and Make a Portage at Lockert's Shoals in Broad River," enacted December 18, 1813. The commissioners were Captain Joseph Hughes, Major James Ming, and Captain Samuel Hancett. McCord, ed., *The Statutes at Large of South Carolina,* 7:582.

CHAPTER 4. *Building the South Carolina Canal and River-Navigation System, 1817–1828*

1. David Rogerson Williams (1776–1830), soldier, manufacturer, congressman, and governor of South Carolina, was born at Robbin's Neck, South Carolina, and educated at Rhode Island College (Brown University). He edited and published the *Charleston City Gazette and Weekly Carolina Gazette* from 1801 to 1804 and was a planter and cotton manufacturer near Society Hill, South Carolina, from 1803 until his death. In 1806 he married Sarah Power. He married a second time, in 1809, to Elizabeth Witherspoon. He was a member of Congress in 1805–9 and 1811–13. In 1813–1814 he was a brigadier general in the U.S. Army. He served as governor of South Carolina in 1814–16. Interested in agricultural matters, he became an early southern manufacturer of cotton products. He was an active leader in South Carolina's internal improvement movement. He was accidentally killed in November 1830 while superintending the construction of a new bridge over Lynches River, Witherspoons Ferry, on the road to Georgetown in Georgetown County, South Carolina. See Kohn, *Internal Improvement,* 600; Sobel and Raimo, eds., *Biographical Directory of the Governors of the United States, 1789–1978,* 1394; and *Biographical Directory of the United States Congress,* http://bioguide.congress.gov (accessed April 30, 2009).

2. Andrew Pickens Jr. (1779–1838), soldier, planter, and governor of South Carolina, was born in Edgefield County, South Carolina, the son of Revolutionary War general Andrew Pickens. The son was educated at the College of New Jersey (Princeton University). He was twice married, first to Susannah Smith Wilkinson in 1804 and later to Mary Willing Nelson. He served as a lieutenant colonel in the U.S. Army during the War of 1812 and as governor of South Carolina in 1816–18. In 1819 he relocated to Alabama to engage in cotton planting. See Sobel and Raimo, eds., *Biographical Directory of the Governors of the United States,* 1394–95.

3. McCord, ed., *The Statutes at Large of South Carolina,* 6:91–92, quoted in Phillips, *A History of Transportation in the Eastern Cotton Belt to 1860,* 83.

4. In 1820 the Board of Public Works reported, "In North Carolina measures are taking to remove the obstruction in that state, and extend this navigation [the Broad River] forty miles above the line of the two states, near to the foot of the Blue Ridge." Board of Public Works, *Plans and Progress of Internal Improvement in South Carolina,* reprinted in Kohn, *Internal Improvement,* 75 (hereafter, *Plans and Progress*).

5. Also in 1820 the Board of Public Works reported: "When this river [the Catawba River] is opened, boats may ascend to a place called the Pleasant Gardens above Morgantown, and three hundred miles from Charleston by land. This navigation will enter the Blue Ridge, passing through the finest counties of North Carolina, and approaching within fifty miles of the navigable waters of the Holston in Tennessee" (Board of Public Works, *Plans and Progress,* reprinted in Kohn, *Internal Improvement,* 74–75). Despite these optimistic plans, very little was done to improve the rivers for navigation in North Carolina. A company for improving the Catawba River within that state was incorporated prior to 1800, but it was inactive and was forced to forfeit its charter. See Weaver, *Internal Improvements in North Carolina,* 65.

6. South Carolina's Great Pee Dee River is called the Yadkin River in North Carolina. By 1818 the state of North Carolina had chartered the Yadkin Navigation Company to open the Yadkin River for navigation from Wilkes Court House in the western part of North Carolina to the mouth of Westfields Creek, above Cheraw—a distance of 247½ miles. See *Southern Patriot,* November 5, 1818, 3, and February 15, 1819, 2. Like other optimistic proposals for improvements in North Carolina, such as those for the Broad and Catawba rivers, very little work was done to improve the Yadkin River for navigation.

7. Letter to the editor, signed "A South Carolinian," *Southern Patriot,* November 12, 1819, 2.

8. Hollis, "Costly Delusion," 29–43.

9. "The Board of Public Works, has, within a few days past, been deprived of one of its most useful members, by the death of Gen. W. R. Davie" (*Report of the Board of Public Works to the Legislature of South-Carolina for the Year, 1820,* reprinted in Kohn, *Internal Improvement,* 23).

10. "Public Works," *Southern Patriot,* February 4, 1820, 2.

11. William Jones Myddleton (1781?-1837?), planter, was born in South Carolina, lived in St. Matthew Parish, and married his cousin, Mary Hart. During the War of 1812 Myddleton served in the state militia as a captain and later as a colonel. He served in the South Carolina Senate as representative of St. Matthews in 1814–15, 1816–17, 1818–19, and 1820–21. In 1820 he was contractor to the Board of Public Works for clearing the Santee and Congaree rivers, and in 1821–23 he served on the Board of Public Works. In 1824 he became president of the Richland Steam Boat Company. Myddleton became president of the Columbia branch of the Bank of the State of South Carolina, ca. 1825–26, and discount clerk, ca. 1828–36. See Bailey, Morgan, and Taylor, *Biographical Directory of the South Carolina Senate,* 2:1184–85; and Kohn, *Internal Improvement,* 601.

12. Members of the board served without salary. In 1820 the staffs' yearly salaries were as follows:

John Wilson	$3,500
Thomas Baker	$2,500
Thadeus Sobieski	$2,000
John Couty	$1,500
W. M. Gibbes	$1,000

(*Southern Patriot,* February 4, 1820, 2).

13. Kohn, *Internal Improvement,* xiii.

14. Board of Public Works, *Plans and Progress* (1820), reprinted in Kohn, *Internal Improvement,* 71.

15. *Report of the Civil and Military Engineer for 1818,* reprinted in ibid., A16.

16. Ibid.

17. Ibid., A17.

18. Mills, *Inland Navigation.*

19. Introduction to ibid., in Kohn, *Internal Improvement,* 140, from the *Charleston Courier,* December 12, 1821. The resolution in the Robert Mills paper at the Library of Congress is different. It reads: "Resolved that the Commissioners on internal improvements be directed to have an accurate survey made from Granby to Charleston of the land on which a canal and road is recommended as pr. plat furnished by Mr. Mills, and that they do report to the next session of the Legislature a plat thereof, designating the particular lands or Plantations with names thereof through which it will pass, the state of soils [indistinct] with an estimate of the expence or probable cost and the most eligible mode of commencing the [indistinct] work" (motion of P. Moses, no date. Robert Mills Papers, Library of Congress, number 0861).

20. *Report of the Board of Public Works for 1821,* reprinted in Kohn, *Internal Improvement,* 140.

21. *Report of the Superintendent of Public Works for 1823,* reprinted in ibid., 309.

22. Bryan, *Robert Mills,* 152.

23. Granby functioned as the river port city for Columbia. Abandoned in the nineteenth century, it was located on the west bank of the Congaree River immediately below present-day Cayce, a suburb of Columbia.

24. Hamburg was located across the Savannah River from Augusta, Georgia. Like that other South Carolina steamboat destination, Granby, Hamburg no longer exists. Both were sited on low lands subject to periodic flooding, and both were abandoned in the nineteenth century.

25. *Report of the Civil and Military Engineer for 1819,* reprinted in Kohn, *Internal Improvement,* 5.

26. Ibid., 6.

27. *Report of the Superintendent of Public Works for 1824,* reprinted in ibid., 344. To correct this error, Blanding entered into a contract with Mr. Charles McCulloch: "To give this work [the Catawba Canal] complete protection against any future rise of water in the river, it became necessary to raise the masonry of the guard lock three feet

six inches higher, and to give the banks of the upper part of the canal a corresponding elevation. This Mr. McCulloch has contracted to perform for the sum of $830.80" (ibid., 344).

28. Robert Mills, *Statistics of South Carolina,* 157–58.

29. The Congaree and Santee Steam Boat Company was established in 1818. Its first steamboat was the *Columbia,* which began operations from Charleston to Granby on the Santee and Congaree rivers in 1819. Unfortunately on its maiden voyage to Granby, the *Columbia* with its two tow boats went aground near Lanud's Ferry on the Santee River. Because of the low level of water on the Santee, owing to the drought of 1819, she remained aground for several days. One observer reported, "It is much regretted that on her first voyage, that [the] Santee River should prove so exceedingly low, (as low as I ever knew it)" (*Southern Patriot,* July 3, 1819, 3). Initially steamboats traveled from the mouth of the Santee—usually the North Santee which was deeper than the South Santee—to Charleston via the ocean. By the end of 1821, under the superintendence of R. T. Morrison, the Board of Public Works improved Sullivan's Island narrows, Price's Inlet, and Bull's Island narrows, thereby allowing an inland passage of steamboats from the mouth of the Santee to Charleston. See *Report of the Board of Public Works for 1822,* reprinted in Kohn, *Internal Improvement,* 170.

30. *Southern Patriot,* May 21, 1819, 3.

31. *Report of the Civil and Military Engineer for 1818,* reprinted in Kohn, *Internal Improvement,* A10. These individuals included General Wade Hampton, Colonel Hampton, Colonel Thomas Taylor, Colonel John Taylor, Major Thomas Taylor, and Jesse, Benjamin, and James Taylor.

32. *Report of the Civil and Military Engineer for 1819,* reprinted in Kohn, *Internal Improvement,* 11.

33. *Report of the Superintendent of Public Works for 1823,* reprinted in ibid., 280.

34. *Report of the Civil and Military Engineer for 1818,* reprinted in ibid., A10.

35. Murray's Ferry, later called Gourdin's Ferry, was northeast of Pineville on the Santee River.

36. Board of Public Works, *Plans and Progress* (1820), reprinted in Kohn, *Internal Improvement,* 72–73. This report was widely circulated and subsequently republished in the United States. See, for example, *American Farmer, Rural Economy, Internal Improvements, Prices Current* (Baltimore), December 1, 1820.

37. Colonel Myddleton described for the Board of Public Works the work he and his men undertook in 1820 on the Santee River:

> Inclosed you will receive rough drawn sketches of the upper and lower falls in Santee river. The upper falls are occasioned by a ledge of rocks extending down the river, on the western shore, for about two hundred yards, and come to a point within twenty feet of the eastern shore, opposite the angle of the river at A. These falls were examined by me, the latter end of May last [1820], when the water was very low. The rocks in the middle of the river, were from six inches to two feet above the surface of the water, leaving two channels, the one on the western shore averaging about two feet water over a rocky bottom, for the distance of two hundred

yards. The channel on the eastern shore, was quite as shallow, but only seventy yards in length, with a sandy bottom. After having examined this channel, and finding no rock, the situation of these falls, was reported to you verbally, when I received your instructions, to cut away the point of land at A. so as to widen and deepen the channel. After clearing the point of timber, a ditch twelve feet wide was cut across, extending to deep water above and below, leaving a space of fifty feet between the ditch and the river, at the extreme point. The ditch was sunk about four feet below the surface of the water, and the point between the ditch and the river cut away nearly to the water level. A small rise in the river prevented any further operation. It may be found necessary, at very low water, to put in a jettee extending from the western shore at B. so as to throw a sufficient current through this channel, to keep it always open. The lower falls are about four and a half miles by water, lower down the river, and are more difficult. They extend nearly a quarter of a mile, and cover nearly the whole bed of the river, leaving a very narrow shoal and imperfect channel, on the eastern shore. The river having risen five or six feet at the time of the examination, it was impracticable to make a very correct survey of the falls. I feel confident, however, from the view I had of these falls and the situation of the neck of land, on the eastern shores, that the cheapest and most certain method of forming a safe and durable channel, is to cut a canal across the point, from A. to B. There is a bank of from six to twelve feet high above low water mark, and about twenty feet wide next to the river, all the way round the bend; the intervening space is very low, about level with low water, with the exception of two small ridges of fifteen or twenty feet wide and three above the level of the low ground. (letter from W. J. Myddleton to A. Blanding, November 23, 1820, in Board of Public Works, *Plans and Progress* [1820], reprinted in Kohn, *Internal Improvement,* 51).

38. Myddleton's instructions for work on the Congaree and Santee rivers were as follows:

The Work will commence on the Congaree at the point where Mr. Barbour left off, or where Coll. Myddelton shall find that the first work is necessary; He will proceed down to the Mouth of that River & then remove to such part of the Santee as most requires clearing. One white Overseer ought to be engaged on each Sett of Flatts, & about fifteen slaves; but where occasion requires, the number may be increased to twenty. Weekly supplies of provisions ought to be delivered to the Overseers, and the rations to the Men delivered out daily. One Cook may be allowed to each sett of Flatts. The Overseer must be required to make a weekly return of the rations delivered out each day, & an acct. of all Tools received, lost, or damaged. Overseers not complying with this regulation must be discharged. In disposing of the Logs removed or taken up, such as will sink must be freed from all their Snags, cut into small pieces, and disposed of in deepwater & out of the Channel, such as float, must be carried to the land & placed above high water mark; or burnt, or cut into pieces & disposed of in eddies or Old Rivers. All the Trees that are likely soon to fall into the River must be cut down & disposed of.

> Hanging Timbers are to be removed & such Trees as are likely hereafter to become dangerous are to be girdled. And this work must be particularly attended to when the state of the River will not admit of working in its channel. Whenever the state of the River will admit of removing the ledge of Rocks in the Santee, Coll. Myddelton will begin that work, & if Blasters are necessary they shall be sent him. He will also remove the ledge of Rocks on the Congaree near Belville, & the ledge below Granby. Coll. Myddelton will in no case cut through a Neck but without Orders from the Board, but he will examine & report any he thinks ought to be cut. In carrying on the work Coll. Myddelton will take care to cause as little injury as possible to the Inhabitants on the River, and he will stipulate with the persons from whom he hires Slaves, that in case of their becoming ungovernable or committing outrages on the property of the Inhabitants, he shall be at liberty to discharge them. In the Contracts with Overseers it shall be stipulated that they shall attend the men under their charge & oversee them when removed to the Road. The Flats must be supplied with the proper medicines and due attention paid to the Sick; medical aid may be demanded when necessary. Deductions for labor shall be made when more than one weeks labor is lost in the Month, or any more favorable arrangement may be made. Slaves in the Flats shall be allowed three quarters of a pound of bacon, or a pound & a quarter of Beef per day, and at the rate of one peck of meal & two quarts of peas per Week, and one gill of Whiskey or molasses per day. The nature of the rations may be changed, provided the expence is not increased. A. Blanding ("Instructions for Coll. Myddelton in carrying on the Work on the Congaree & Santee Rivers," South Carolina Department of Archives and History, Department of Public Works, Miscellaneous Papers, February 20, 1820 [S208006]).

39. Ibid. A gill of whiskey is usually a quarter of a pint (0.118 liter) but may have been half a pint on these works.

40. Letter from Myddleton to Blanding, November 23, 1820, in Board of Public Works, *Plans and Progress* (1820), reprinted in Kohn, *Internal Improvement,* 52.

41. *Report of the Superintendent of Public Works for 1823,* reprinted in ibid., 280.

42. Wilson described the falls in the Congaree River at Columbia: "A short distance above Granby, and above the mouth of Rocky Creek, are the first conspicuous ledges of rock which obstruct the navigation of the Congaree. The extent of these rapids is about half a mile, with a perpendicular rise of eight feet. From thence to Columbia ferry, the current is regular. From this ferry to Young's Mill-dam, at the confluence of the Broad and Saluda rivers, is one continued shoal. The distance is one mile, and the rise fourteen feet six inches" (*Report of the Civil and Military Engineer for 1818,* reprinted in ibid., A10–A11).

43. The 1822 report had a description of the work that remained along the Congaree and Santee rivers:

> Whilst upon the subject of the lower navigation connected with this river [the Santee River], it may be proper to notice the conditions of the upper part of this stream and its branches, and point out those obstructions which require to be

removed to enable the steam boats to pass at low water. The most material point, and which may require the greatest expenditure of labor to improve, is *Chicken creek,* just before you come to the division of the Santee. This creek takes pretty much the general direction of the southern branch of the river, and were it cleaned out, would save a circuit to the boats of about 12 miles; but this is of minor importance, compared with other objects which would be attained, for we should thereby give to the South Santee the same advantages of navigation at low water possessed by the North. When the river is low, the Steam boat is compelled from the want of water to take the circuit of the North river; and the 4 or 6 mile creeks, to get into South Santee again. Were this creek cleaned out, the current, which in low waters chiefly runs into North Santee, would divide, and supply also the South river. The labour, as was stated before, would not be great to effect this desirable object, for the canal boats are now in the habit of passing through Chicken creek.

The next point requiring attention, and which if improved, would facilitate the passage of boats, is where the *Congaree and Wateree rivers meet.* The great bend made by the river below, and which approaches very near again the point of starting, could be avoided by following the direction of Manigault's creek, or Little river opposite to the Wateree, which brings you very soon into the Santee river, thus avoiding a circuit of many miles. The labor is not too great to accomplish this.

Barber's Cut, about 21 miles below Granby, requires some considerable labor to put it into a good state for the passage of the steam boats. The old river is now impassable, and this cut more imperiously demands attention. It was originally made too narrow, and without cutting off the trees upon the margin to provide against the wash of the current, which might have been naturally looked for, in this peculiar situation.

Just below *Pinckney's creek* 9 or 10 miles lower down, some *stumps of trees* have been sawed off shallow, and interrupt the navigation at low water. By improving further, a creek (less than a mile) near Marion's point, now passable for canal boats, a current of 3 miles might be avoided round this point. *Horse creek,* wants a little work done upon it,—as also a cut 5, or 6 miles above Dr. Dwight's.

The last noted obstruction, is a *shifting sand bar,* about 11 miles below Granby. There is at present a tolerable good channel, but to keep it so, a jettee ought to be constructed to divert the current in low waters to pass through it. (*Report of the Board of Public Works for 1822,* reprinted in ibid., 172–73)

44. Board of Public Works, *Plans and Progress* (1820), reprinted in ibid., 73.

45. Wilson wrote: "From the resolution passed by the legislature in December 1817, directing the improvement of the Broad and Saluda rivers, and designating them as the first objects of attention, and being obliged, as before stated, to confine my personal superintendence to one point, I proceeded to examine the first obstructions mentioned in my Report of the last year [1818], under the names of Beard's and Senn's Falls [on the Saluda River, twenty miles upstream from Columbia]" (*Report of the Civil and Military Engineer for 1819,* reprinted in ibid., 11). Wilson was criticized for constructing

canals and river navigations on the upper rivers before the completion of river improvements on the lower rivers, thereby preventing the use of these navigations. He was defended in this decision by a Charleston newspaper: "In our opinion the conduct of the State Engineer required no defence; we know that he had been zealous and indefatigable in carrying into effect the intentions of the Legislature. By a resolution of that body he is expressly directed to open the navigation of the Broad and Saluda rivers, and it was generally understood that he was to commence his operations at that point" (editorial, *Southern Patriot,* May 21, 1819, 3).

46. Wilson described the obstructions in the Saluda River that the Saluda Canal would bypass:

> The first obstruction to the navigation of the Saluda, is at the mouth of that river; a rocky shoal extending three hundred yards with a perpendicular rise of nearly nine feet. From hence the current is regular for one mile and three quarters to Beard's falls. This formidable obstruction is formed by large masses of rocks, extending across the river, and dividing the water into numerous channels, through which boats cannot be navigated. These rapids extend up the river two hundred yards, and the perpendicular rise in that short distance is fourteen feet. At about one mile higher up, at Senns, are shoals extending for a mile and a half up the river; the rise in this distance has been estimated at thirteen and a half feet; but it is not regular, for the current is rapid and broken only at particular places. (*Report of the Civil and Military Engineer for 1818,* reprinted in Kohn, *Internal Improvement,* A12).

47. *Southern Patriot,* March 23, 1819, 2.

48. Wilson described the circumstances of the assignment of the contract initially awarded to Strode for building the Saluda Canal:

> In order to carry on this work, I entered into a contract with Capt. Thomas Strode on the 25th January, 1819, who agreed to furnish the requisite number of mechanics and labourers. This gentleman, according to his contract, placed 120 men upon the works on the Saluda in March, but owing to some unexpected difficulties, was obliged to assign over his Contract on the 22d of August, to certain Creditors who had furnished to him supplies for carrying on the works designated in the original Contract of 25th January 1819. This assignment, which limits the existence of the original Contract to the period of the extinguishment of his debt to the State, for advances made him and the debts due by him to those who furnished the works with materials and provisions, I considered to be reasonable, I therefore agreed to it, under the proviso, that the assignees were to prosecute the work according to the terms and conditions of the Contract of 25th January. (*Report of the Civil and Military Engineer for 1819,* reprinted in Kohn, *Internal Improvement,* 12)

49. Elsewhere in the same report the distance is given as two miles and forty-seven chains and twenty-four links. Two miles and forty-seven chains and twenty-four links equals two miles, 3102 feet, and 15.8 feet or two miles 3117.8 feet (2.59 miles).

50. Board of Public Works, *Plans and Progress* (1820), reprinted in Kohn, *Internal Improvement,* 73.

51. Wilson provided the state legislature with a description of this dam at the upstream end of the Saluda Canal:

> After a careful investigation, it appeared to me most expedient to construct a Dam at a bend of the river above Senn's Falls, and to conduct the navigation by a Canal running along the Eastern margin of the Saluda, and terminating in Broad River, just above the mouth of the Saluda, so as to avoid both Falls. The dam is four feet high, firmly secured to the Rocks, and extends across the river, a distance of 700 feet, with stone abbutments at each extremity of it. It is built of stone and timber. The first parallel lines of square logs are secured to the Rocks by Iron Bolts; these are connected by smaller timbers or ties, dove-tailed into the front log, and fastened into the back by a notch and pin. The space between the timbers when carried to the height required, is masoned in with stone; the top is then graveled and sheeted with good plank: the whole has been executed in a substantial manner. (*Report of the Civil and Military Engineer for 1819,* reprinted in Kohn, *Internal Improvement,* 11–12)

52. *Report of the Superintendent of Public Works for 1826,* reprinted in ibid., 468–69.

53. *Report of the Superintendent of Public Works for 1823,* reprinted in ibid., 304–5.

54. Wilson described the Saluda River navigation upriver (west) of the Saluda Canal: "For ten miles above Senns [the beginning of the Saluda Canal], the navigation of the Saluda is more obstructed by fish-dams than by natural obstacles; the whole rise not exceeding eight feet, and only one shoal occurs in this distance, at Gleckley's Island, with a rise of two and a half feet. The rapids at Caseman's, extend half a mile up the river, the rise being ten feet. The river is navigable for one quarter of a mile above this obstruction, to Drehr's Mill and Ford, where there is a rise of one foot in the first hundred yards; and in the whole extent of the rapids, six hundred yards, the rise is ten and a half feet" (*Report of the Civil and Military Engineer for 1818,* reprinted in ibid., A12).

55. Board of Public Works, *Plans and Progress* (1820), reprinted in ibid., 19.

56. Also reported as a twenty-one foot drop.

57. "Mr. Kibbe has undertaken the works at Drehr's Falls, fifteen miles above Columbia, and has placed a strong force on the works" (*Southern Patriot,* March 10, 1820, 2).

58. *Report of the Superintendent of Public Works for 1828,* reprinted in Kohn, *Internal Improvement,* 305. In their 1820 report the board provided a more complete description of Drehr's Canal:

> At Drehrs, there is a fall of 21 feet, in the distance of 1100 yards. The canal is taken out of the river at the Mills, where a short wing dam will give 2½ feet water at the head of the falls in the driest season. It is extended to the foot of the falls on the same level, where it will enter the river by 3 locks in one chain. The water in freshes rises very high at the top of these falls and a strong guard lock, rising

seven feet above the water level in the canal, was found necessary to give protection to the works below. The canal crosses two ravines and one creek, requiring two culverts and one aqueduct. The whole of this canal has been excavated. One culvert and the aqueduct have been constructed—One occupation bridge, for the use of Drehr's mills, has been erected. The guard lock and dam are now building, and a considerable quantity of cut stone for the works is prepared. A stone building for the Lock-keeper has its walls nearly up. The water that descends the ravine near the head of the works, cannot be discharged without passing very near the bottom of the canal. This space will not receive a stone culvert; one of cast iron has therefore been prepared, and most of the pieces already delivered. The castings for such a culvert, 60 feet long and 22 inches interior diameter, cost $300. A stone culvert costs about the same. (Board of Public Works, *Plans and Progress* (1820), reprinted in Kohn, *Internal Improvement,* 42)

59. Mitchell, *Mitchell's Compendium,* 52–53.
60. "Description of a wooden Lock to be erected at Loricks Mill on Saluda river":

The lock must be put in at the point marked by me when on the ground. The sill of the lower gate must be two feet six inches below the surface of common summer water. The sill of the upper gates must be two feet six inches below the top of Mr. Loricks dam. The top of the lock must be at least six inches [higher] than the top of the gates, and the top of the gates must be as high as the rise of water in the highest freshes. The Chamber of the lock must be seventy feet long and ten feet wide. The recesses for the gates each nine feet long and eighteen inches deep. The head above the upper recess must be eight [?] feet long. The distance below the lower gates to the steps must be twelve feet and then a descent of eight steps each having a tread and rise of one foot. The whole lock will then be 116 feet long. The walls will be built of timber as in the annexed plan and the spaces between them filled with puddle made of good clay and well worked. The timber of the facing of the lock must be of good heart pine one foot square well hewn and jointed down so as to be water tight. The cross ties must be of the same stuff well dovetailed into the front and back timbers. The backing of timber, which will be covered may be of any stuff that will square a foot and may be hewn only on two sides. At the head of the lock, the cross timbers will extend five feet beyond the wall: and entirely across the head there must be plank piling down to the solid and above it three feet of puddle from A to A in the annexed plan. Across the breast of the lock from B to B, it must be piled as also from C to C and from D to D. Within the facing of timber, there must be a sheeting of [indistinct] plank so as to cover all the joints in the timber, and back of this the puddle must be put in. The bottom of the lock will no doubt be in rock, in which the first course of timber must be bolted to it. The gates must be worked with levers and the wickets with rack and pinions. The locks must be filled through wickets in the upper gates and to prevent the water from filling the ascending boats, there must be worked into the wall at A-A timber five feet high and one foot thick, the top of which must be on a level with the sills of the upper gates. The canal must not be cut till the lock bottom is formed, and then the stuff taken from the canal must be used for puddling the lock. The line of canal is marked out. It must be twelve feet wide at bottom and slope on each side eighteen inches in every rise of one foot. The space between C & D should be filled with stone & puddle if the stone can be procured.

A. Blanding

(South Carolina Division of Archives, Department of Public Works, Road, River and Canal Projects File, 1820–1845, May 28, 1824 [S208009])

The plan originally included with these instructions has not been located and is thought to be no longer extant.

61. Board of Public Works, *Plans and Progress* (1820), reprinted in Kohn, *Internal Improvement,* 74.

62. *Report of the Board of Public Works for 1821,* reprinted in ibid., 125.

63. Ibid.

64. "Public Improvements," letter to the editor from "An Unprejudiced Observer," *Southern Patriot,* October 10, 1821, 2. The same individual wrote: "In prosecuting the works, advantage will occasionally be attempted by some of the contractors, as in the works at Drehr's Mills, and without such experience [of the commissioners], the deception might pass unnoticed" (ibid.).

65. The 1823 report stated: "This [Drehr's Canal] overcomes a fall of 21 feet. It is 1100 yards long, on a difficult line. It has one guard and three lifting locks, 1 bridge, 2 ordinary culverts, one of stone, the other of cast iron, 1 aqueduct, 1 bridge, 1 waste wier [weir], and 1 lock-keeper's house of stone. The work was begun in 1820, and finished in 1821. It cost $78,139.63, which has been paid. This canal is an essential link in the navigation of this river, and comes into operation with the Saluda canal" (*Report of the Superintendent of Public Works for 1823,* reprinted in Kohn, *Internal Improvement,* 305).

66. *Report of the Superintendent of Public Works for 1824,* reprinted in ibid., 347.

67. Ibid., 334–35.

68. Ibid., 334.

69. "Saluda Navigation," *Southern Patriot,* March 22, 1825, 2.

70. *Report of the Superintendent of Public Works for 1826,* reprinted in Kohn, *Internal Improvement,* 468.

71. *Report of the Superintendent of Public Works for 1823,* reprinted in ibid., 305–7.

72. The length of the Saluda River navigation varied with different reports. In the 1826 report, Blanding claimed that the Saluda River navigation extended one hundred miles above Lorick's Mills (estimated to be fifty water miles above Columbia)—for a total distance of 150 miles instead of 118 miles. See *Report of the Superintendent of Public Works for 1826,* reprinted in ibid., 469.

73. Ibid., 468.

74. Ibid.

75. *Report of the Superintendent of Public Works for 1823,* reprinted in ibid., 308.

76. Blanding explained the need for such an expensive dam across the Broad River to the legislature:

> At this river lock No. 4 [on the Saluda Canal where it enters the Broad River], was to be constructed so as to have two and a half feet water in it, after a dam should be put in, across Broad river, six feet high. With the aid of this dam, boats from the Saluda canal, will enter Broad river, cross it, and enter the Columbia canal, which it was proposed to extend on the same level, to Granby. This dam, having an elevation equal to all the fall between it and Guignard's island [on the Broad River, 1½ miles above the confluence of the Broad and Saluda rivers], will of course cover that fall, and supercede any labour on that part of the river. Between this island and the main, there is a natural canal extending the whole distance of the falls of the Bull sluice. A lock of eight feet lift, constructed in this canal, will gain the still water at the head of these falls.... the dam is essential to it [the plan to pass canal boats from the Saluda Canal to the Columbia Canal across the Broad River and to pass canal boats down the Bull sluice to the entrance of the Columbia Canal]. Without it, the work would be less perfect, it would require twelve feet more lockage and one and a half mile of canal: for, without the dam, lock No. 4, in the Saluda canal, would be six feet above the water in Broad river, and another lock of six feet lift would be there required. And when, in Broad river, at this point, a boat would be six feet below the Columbia canal, to which, it could ascend only by a lock of that lift. Without the dam, the fall between the mouth of Broad river and Guignard's island, must have been overcome by a canal through ground embarrassed by many difficulties. The line would encounter a most precipitous hill of solid granite, that rises immediately from the water's edge. Several ravines must be passed requiring walls and culverts. It is rare, that more difficulties are thrown in a space of one and a half mile. Without the dam, the navigation would be less perfect, less safe. For then the boats from the Saluda canal must enter the Broad river, at a rapid, and the Columbia canal could be gained neither with ease nor safety. (Board of Public Works, *Plans and Progress* [1820], reprinted in ibid., 39–40).

77. Puddling was used extensively on the canals of South Carolina. Blanding reported, "Where puddle has been properly used, (and it has been uniformly used on all canals I have directed, which have not been cut entirely within soil,) we have no leaks but what have been stopped at a trifling expense, and no new ones appear" (*Report of the Superintendent of Public Works for 1823*, reprinted in ibid., 308).

78. Ibid., 303.

79. Board of Public Works, *Plans and Progress* (1820), reprinted in ibid., 42.

80. *Southern Patriot,* March 10, 1820, 2.

81. *Report of the Superintendent of Public Works for 1823,* reprinted in Kohn, *Internal Improvement,* 303. Quoting a letter sent from Columbia on July 29, 1823, the August 1 issue of the *Southern Patriot* provided additional details of the death of contractor McKenzie (often spelled *McKinzie* or *M'Kinzie*) on the Broad River dam, which was located immediately above Columbia:

We are sorry to state that Mr. M'Kinzie, late contractor for erecting a dam across the river opposite this town, at the mouth of the Saluda river, was drowned on Saturday evening last, in the act of crossing the river near said dam:—It appears that near the close of the day Mr. M'Kinzie, and two of his workmen, one by the name of Milton Ladd, and the other named Patrick Dunn, went over the river from the Columbia side, to see if no part of the dam was giving away, as there was a considerable swell in the river; that they continued in the river near some islands until dark, when they set out to re-cross the river, and in the act of doing so, were carried over the dam. Mr. Dunn states that they were not apprised of their danger until within a few feet of the dam, where the current is so rapid as immediately to swallow up every thing that comes within its influence, when Mr. M'Kinzie rose up in the boat, and observed, "we are all lost," upon which Mr. Dunn immediately leaped into the water, and was carried over the dam, and continued for some time in the surge, rising and sinking, and dashing alternately against the dam and the rocks until he was nearly exhausted, when he fortunately disengaged himself from the whirlpool, and got on an island where he remained until Sunday morning about 10 o'clock. Mr. M'Kinzie and Mr. Ladd have not been heard of since; and are supposed to have been immediately drowned.

Mr. M'Kinzie was an excellent mechanic, a steady, honest, and industrious citizen. His loss is regretted by all who know him.—*Gazette*

82. *Report of the Superintendent of Public Works for 1824*, reprinted in Kohn, *Internal Improvement*, 348.

83. Ibid.

84. *Report of the Superintendent of Public Works for 1826*, reprinted in ibid., 469–70.

85. Blanding reported to the state legislature:

> During the three first quarters of the present year [1826], 18,968 bales of cotton passed through it [the Columbia Canal], loaded on board of 230 boats, and during the same time there were 221 other boats which passed through this canal. The whole tolls in that period amounted to $1303.77 charging only four cents on each bale, and other articles in proportion. It will thus be seen that it is of great importance that there should be no interruption in this navigation. Had the season not been an unusually dry one, or had the dams at the head of this canal been sufficiently tight, the transportation would have been much greater, and the income considerably encreased. If there were no failure from want of water, I believe that more than 30,000 bales of cotton would annually pass this canal. (Ibid., 470)

86. Blanding reported contracting with Colonel Moman for his services and his boat to investigate the Saluda River: "I engaged Col. Moman, a gentleman well acquainted with the navigation of Broad river, and in whose judgment I had great confidence to furnish me with one of the largest boats used on that river [the Broad], manned with 5 experienced boatmen. This boat is 54 feet long, 8 feet wide, and draws 18 inches water when loaded with 60 bales of cotton, her ordinary burthen" (*Report of the Superintendent of Public Works for 1823*, reprinted in ibid., 305).

87. Board of Public Works, *Plans and Progress* (1820), reprinted in ibid., 41.

88. *Report of the Board of Public Works for 1821,* reprinted in ibid., 123.

89. *Report of the Superintendent of Public Works for 1823,* reprinted in ibid., 303.

90. *Report of the Civil and Military Engineer for 1818,* reprinted in ibid., A11.

91. Ibid.

92. Wilson reported to the state legislature: "In June labourers and a few mechanics were placed on the Broad river, at Rive's and Wade's mill, for the improvement of the Sluices: they ascended with their operations on this river, to Chappell's mill, which is about eight miles below the line designated in the contract, and the point where it was intended the Contractor should discontinue to improve the Sluices" (*Report of the Civil and Military Engineer for 1819,* reprinted in ibid., 11).

93. Wilson reported to the state legislature: "On the 31st of August, as I had reason to be dissatisfied with the manner in which these works had been conducted, the Contractor was directed to suspend the work on Broad river" (ibid.).

94. Board of Public Works, *Plans and Progress* (1820), reprinted in ibid., 19.

95. Ibid.

96. "An Act Appointing Commissioners to Lay Out and Make a Portage at Lockert's Shoals in Broads River," enacted December 18, 1813. The commissioners were Captain Joseph Hughes, Major James Ming, and Captain Samuel Hancett. See McCord, ed., *The Statutes at Large of South Carolina,* 7:582.

97. Board of Public Works, *Plans and Progress* (1820), reprinted in Kohn, *Internal Improvement,* 46; March 1820 newspaper clipping from a Columbia, South Carolina, newspaper, reprinted in Kohn, *Internal Improvement,* 67; *Southern Patriot,* March 10, 1820, 2.

98. Two different lengths of this canal were reported to the state legislature. The 1823 report stated that its length was one mile and thirty-six chains, or one mile and 2376 feet (that is, 1.45 miles versus 1.38 miles reported in 1820). See *Report of the Superintendent of Public Works for 1823,* reprinted in Kohn, *Internal Improvement,* 302.

99. Board of Public Works, *Plans and Progress* (1820), reprinted in ibid., 45.

100. Ibid.

101. Ibid., 46.

102. Ibid.

103. *Report of the Board of Public Works for 1821,* reprinted in ibid., 123.

104. *Report of the Board of Public Works for 1822,* reprinted in ibid.,164–65.

105. The canal was completed early in 1824 but was not immediately accepted. Blanding explained to the legislature:

> This canal [Lockhart's Canal] was finished early this year [1824], but has not been received. The commissioners appointed on it, have not taken it under their charge. Having confidence that there was no imperfection in the work, unless it should be found in the two upper locks, I made a settlement for the work and paid the contractors all but $3731.49, reserving that sum till the work should be fully tested. I have not been able to examine it this autumn, so that the work has not been received. I understand that the September fresh, was completely

excluded from this canal, by the protection afforded by the Guard lock, so that it received no other injury, than that the gates of the two lower locks, which were entirely under water, floated from their fastenings. Orders have been given to replace and secure them against similar accidents. (*Report of the Superintendent of Public Works for 1824,* reprinted in ibid., 346)

106. Ibid., 347.

107. Board of Public Works, *Plans and Progress* (1820), reprinted in ibid., 74.

108. *Report of the Board of Public Works for 1821,* reprinted in ibid., 123.

109. Ibid.

110. *Report of the Board of Public Works for 1822,* reprinted in ibid., 164.

111. Ibid.

112. *Report of the Superintendent of Public Works for 1823,* reprinted in ibid., 304.

113. *Report of the Superintendent of Public Works for 1824,* reprinted in ibid., 348.

114. *Report of the Superintendent of Public Works for 1823,* reprinted in ibid., 303–4.

115. Initially, in 1823, Blanding reported the cost of the three structures as $199,485.09. The following year he reported that he had discovered an error of $3,702.22 in favor of the estate of the contractor, Mr. McKenzie. See ibid., 304 and *Report of the Superintendent of Public Works for 1824,* reprinted in ibid., 349.

116. *Report of the Civil and Military Engineer for 1818,* reprinted in ibid., A9.

117. *Report of the Superintendent of Public Works for 1823,* reprinted in ibid., 281.

118. Wilson reported to the legislature: "On the 25th January 1819, a contract was entered into with Mr. Mordecai Barbour of the State of Virginia, for the removal of the Logs and Sand which had accumulated in the Wateree and Congaree rivers: and for the formation of such works as might be directed for the improvement of the Broad river, from the mouth of that river to the point where it is intersected by the Newberry and Lexington district lines" (*Report of the Civil and Military Engineer for 1819,* reprinted in ibid., 10).

119. Ibid., 10–11.

120. Letter from Colonel William Nixon to Abram Blanding, November 23, 1820, as printed in Board of Public Works, *Plans and Progress* (1820), reprinted in ibid., 60–61. Blanding explained the situation with the "raft" to the state legislature: "The most difficult points are at Silver Gut, and at Bentley river. There are new channels formed by the water breaking over narrow necks and leaving the old beds, which have been so much filled up as to be impassable in low water. As the lands were in woods of a very heavy growth, when the new rivers were formed, the whole timber growing on the ground they now occupy, has sunk into the channel and is remarkably difficult to remove" (*Report of the Superintendent of Public Works for 1824,* reprinted in ibid., 341).

121. *Report of the Civil and Military Engineer for 1818,* reprinted in ibid., 11.

122. *Southern Patriot,* March 10, 1820, 2.

123. The instructions Blanding gave to Colonel Nixon for clearing the Wateree River were very similar to those he gave to Colonel W. J. Myddleton for clearing the Congaree and Santee rivers, but in Colonel Nixon's instructions, Blanding specified

that Nixon was to begin at the Camden Boat Yard and proceed downriver to the Santee. See "Instructions for Col. Wm. Nixon in Conducting the Work on the Wateree River," South Carolina Department of Archives and History, Department of Public Works, Miscellaneous Papers, February 17, 1820 (S208006).

124. Board of Public Works, *Plans and Progress* (1820), reprinted in Kohn, *Internal Improvement*, 60.

125. Ibid., 21.

126. Letter from Colonel William Nixon to Abram Blanding, November 23, 1820, as printed in the Board of Public Works, *Plans and Progress* (1820), reprinted in ibid., 61.

127. Ibid.

128. Nixon reported to Blanding: "The hands, that I could not discharge during the sickly months, were at work on the road, from Camden to the boat landing, which has been finished, except for about three hundred yards, which remains unfinished, but is passable. This road is now used altogether. Where it is completed, it is handsomely done" (ibid.).

129. Board of Public Works, *Plans and Progress* (1820), reprinted in ibid., 72.

130. Blanding reported to the state legislature at the end of 1823: "In 1819, the lower part of this river was cleared under Mr. Barbour, by the force employed on the Congaree, . . . The great obstruction to the navigation of this river was occasioned by a raft at the entrance of [the] old river, opposite to Manchester. This is now avoided by opening Bentley river, which has been cleared of most of its obstructions. But there is yet in Bentley, work required to render its rapid current more safe. From Bentley to the upper part of Silver Gut, 14 miles from Camden, no work has been done, and this part of it is much obstructed by logs. From Silver Gut to Graves' ford, 28 miles, the river has been well cleared of logs by a force employed on it in 1820" (*Report of the Superintendent of Public Works for 1823*, reprinted in ibid., 281). Bentley River was not a true river but rather a new channel for the Wateree.

131. Blanding reported on the scope of the contract to the state legislature:

> For eight thousand dollars, I made a contract with Messrs. Nixson [Nixon] and McRa, to open a channel twenty-four feet wide at the bottom, and having three feet at the lowest state of the river, through the gravel shoal at Ancrums, and the rock shoal between Pine bluff landing and Cook's mount; and to clear the river of logs, snags and stumps, from the Camden boat landing, to the lower end of Bently river, forming a clear clearance at least sixty feet wide and three feet deep at the lowest state of the river. The contractors also agreed to remove all the large trees standing near it. They also agreed to clear off, or girdle all the trees leaning over the channel, and girdle all the trees growing at Cantey's and Peter's neck's, where the river is likely to break through and form new channels. (*Report of the Superintendent of Public Works for 1824*, reprinted in ibid., 341)

132. Ibid.

133. *Report of the Superintendent of Public Works for 1825*, reprinted in ibid., 428.

134. Board of Public Works, *Plans and Progress* (1820), reprinted in ibid., 52.

135. Ibid., 55–56. Blanding described the Wateree Canal for the state legislature:

From Jones' [mills] a canal is laid out on the same level about three miles and runs over good ground. The depth of digging will average five feet. At the head a guard lock is to be erected rising six feet above the level of the water in the canal. It is supposed, that this elevation will give perfect protection to the work, as the highest fresh of this year rose four feet only at this point. The continuation of this canal will pass Sawney's creek at a point where its high banks afford every facility for erecting an aqueduct, by which the waters of this creek will pass under the canal. Here a new level must be taken by a descent of three locks and the canal extended to the foot of Groves' [Graves'] shoals where it will enter the river by three more locks. The only difficulties in this whole line, of any serious magnitude, are presented in the aqueduct at Sawney's creek, and in the deep digging for about two hundred yards, which must be encountered at a point where a steep hill shuts in near to the water's edge. (Board of Public Works, *Plans and Progress* [1820], reprinted in ibid., 55)

136. Ibid., 56.

137. Blanding explained the problem of the lack of stone to the state legislature: "Diligent and extensive searches were made for good stone in this neighborhood, but none was found nearer than four miles from the river. I have, therefore, directed the locks and culverts to be built of brick, except that, cut stone is to be procured for the hollow quoins, the counter arches on which the gates close, and the breasts of the locks. The expense of constructing locks with these materials is less than with stone, but the work may not be as durable" (ibid.). The brick culverts proved to be defective because of the brick used, and after his experience with them at the Wateree Canal, Blanding never used brick on other substantial works, where he could avoid it: "The last item in the above statement, arose from the work done in taking up and recovering with stone the brick culverts on the Wateree canal which had been built by David Grafton; and which were supposed to owe their defects to the imperfection of the work; and therefore a sum was retained till the work should be taken up, and examined. This was done and the fault was found to be in the material used, and not in the workmanship of it. I therefore paid the [e]state of Grafton, and Messrs. Clark and Cleverly's bill forms a proper charge against the state. Since I have had the sole direction of your works, I have used no brick, and this failure has confirmed me in the propriety of excluding them from all hydraulic structures of importance, where it is possible to procure stone" (*Report of the Superintendent of Public Works for 1826,* reprinted in ibid., 467).

138. Blanding's instructions to Izard read in part: "Your particular attention will be directed to the mortar and cement. All the sand must be sharp and well washed so as to be entirely free from dirt or clay. The roman cement must be such as sets hard and if that which the contractor have now on hand is not good, they must cease using it and procure better" (letter from Abram Blanding to Walter Izard, January 17, 1825, South Carolina Department of Archives and History, General Correspondence of the Superintendent and Board, 1820–1848, [S208004]). "Roman cement," also called hydraulic cement, would resist dissolution in water.

139. *Report of the Board of Public Works for 1821,* reprinted in Kohn, *Internal Improvement,* 122.

140. *Report of the Board of Public Works for 1822,* reprinted in ibid., 163.

141. Blanding reported: "But the death of the contractor early in the summer suspended this work, and some part of the canalling remains in an unfinished state. The principal culverts were finished and only four small ones remained to be put in, when the work was suspended. The materials for these are collected. The guard lock had been undertaken with most of the material on the ground, but the same cause prevented its completion" (ibid.).

142. Ibid.

143. Ibid., 164.

144. *Report of the Superintendent of Public Works for 1823,* reprinted in ibid., 292.

145. Ibid.

146. Ibid.

147. Ibid., 301. Blanding reported on the advantages and disadvantages of each route: "The advantages of this plan [the second plan] are, first, its cheapness, and second, the permanency of the 5 locks above the entrance. The disadvantages are, the imperfection of the navigation, by entering the river before the whole falls are passed, which the first route avoids, and the want of durability in the wooden locks. The advantages speak for themselves. The disadvantages must be considered. They are not as great as at first might be supposed. The navigation from the entrance of the canal into the river to the foot of Gravesford, aided by the two river locks, will be as good as in many parts of the river above, after all efforts are made to improve them. As to the durability of the river locks constructed in wood, I believe that in situations exposed to freshes and rafts of wood, they would last as long as brick. At both Starks shoals and Landsford, there never would be more than 5 feet of the locks, but what might be kept constantly wet. These five feet would be subject to decay" (ibid.).

148. Ibid.

149. *Report of the Superintendent of Public Works for 1825,* reprinted in ibid., 426.

150. Ibid.

151. *Southern Patriot,* January 26, 1826, 2.

152. *Report of the Civil and Military Engineer for 1818,* reprinted in Kohn, *Internal Improvement,* A9–A10.

153. Wilson reported to the legislature: "I regret that I am prevented from presenting to the legislature an accurate survey of these rapids and falls [on the Catawba River]; the plans and draughts of this river being still in possession of the Catawba Company" (ibid., A10).

154. *Report of the Civil and Military Engineer for 1819,* reprinted in ibid., 23.

155. *Report of the Civil and Military Engineer for 1818,* reprinted in ibid., A10.

156. Board of Public Works, *Plans and Progress* (1820), reprinted in ibid., 63. Distance conversion calculated at 1 rod equals 16.5 feet and 1 link equals 0.66 feet.

157. John Couty, assistant engineer to Blanding, provided the legislature a description of this canal, section 1:

> From Powell's Landing to the mouth of Fishing Creek, the distance is 2 ¼ miles 35 roads [rods] and 14 links [that is, 2.25 miles and 586.7 feet or 2.36 miles]. The fall is 50 feet, 5 inches. A dam is required to be built 130 yards long, 2 feet high, extending obliquely in the river to overflow some rocks at the mouth of the canal. The canal to be taken out 170 yards below the present landing, and preserved upon a level along the margin of the river through a red clay soil, well adapted for canaling. The purpose of continuing this level, is to effect a communication with the valley of Fishing Creek, which is better calculated for the contemplated work, than to continue along the margin of the river, which is extremely rocky, and works if constructed there, would be exposed to danger by high freshes of the river; by continuing the same level, lessens the depth of cutting in the dividing ridge in Hicklin's land, which averages 15 feet for 212 yards thence continuing upon the same level extending round the ridge, and then running parallel with Fishing Creek to the place where it will be necessary to lock down into the creek. (Ibid.)

Distance conversion calculated at 1 rod equals 16.5 feet and 1 link equals 0.66 feet.

158. Couty provided the legislature a description of this canal, section 2:

> From Fishing creek to Gill's old mill, the distance is 1–½ mile, 15 rods and 3 links [that is, 1.5 miles and 250 feet, or 1.55 miles]. A dam extended from the old mill place to Mountain Island, 300 yards long, averaging 11 feet high, raises the water sufficiently to overcome all obstacles from that place to the mouth of Fishing creek. By this mode, we avoid the necessity of carrying the line of the canal along the margin of the river, which would in any other way, have been unavoidable, over ground extremely unfavourable by the large masses of rock extending a considerable distance, and in some places projecting into the river. This dam also answers an important purpose in the section below it, the depth of digging in the ridges which the canal must cross, is diminished in proportion to the elevation of the water, and all the works below will receive an ample and certain supply of water from the dam. (Ibid., 64)

Distance conversion calculated at 1 rod equals 16.5 feet and 1 link equals 0.66 feet.

159. Couty provided the legislature a description of this canal, section 3:

> From Gill's mill to Rocky creek the distance is two miles, eleven rods, and six links [that is, two miles and 185.5 feet, or 2.04 miles], fall eighty-nine feet nine inches. The entrance of the canal has to be effected by blasting of rock the distance of one hundred and seventy yards long, averaging eight feet five inches deep, then it enters a valley which may be converted into a basin by building a dam fifty yards long averaging sixteen feet high, which will open the navigation the distance of six hundred yards without digging, and which will be of importance for various purposes. The ridge at the establishment [Mount Dearborn] is two hundred and ninety-seven yards across and the digging will average eleven feet and one inch in depth. It is a red clay soil. The canal is continued upon the same level to about one hundred yards below a brick kill, here it becomes necessary to lock down by

> three locks of eight feet three inches each into a basin created by erecting a dam at a very eligible site, then placing a lock of eight feet three inches lift below the dam. Thence continuing the canal upon a level to where it will be necessary to communicate with Rocky creek. The last mentioned level crosses a ridge with common depth of digging and avoids a very considerable mass or crag of rock. It is also necessary to avoid entering the creek higher up than the point marked on the plan, as the whole distance from the mouth of the ravine which descends from the establishment to the above point, is crowded with large rock, with a very rapid fall and the creek rises here to a great height after heavy rains, sufficient dams would only be built with much labor, and maintained with great expense. Five culverts or aqueducts will have to be built. The soil on which the canal will pass, is a red clay very suitable for canalling. Seven locks of eight feet lift each are required to descend into Rocky creek, below the third lock a basin may be easily effected by building a dam, and which would be essential for giving a sufficiency of water for the four lower locks. (Ibid.)

Distance conversion calculated at 1 rod equals 16.5 feet and 1 link equals 0.66 feet.

160. Couty provided the legislature a description of this canal, section 4:

> From where the canal communicates with Rocky creek to Kingsberry's ferry is two miles, twenty-three rods, twenty-one links [that is, two miles and 394 feet, or 2.07 miles], fall thirty-five feet, two inches, three tenths. The rocky surface of the left bank of the creek prevented the canal from being extended upon it farther, at a moderate expense. Rocky creek may be improved at a small expense to near its mouth. A dam four feet eight inches high and a lock of ten feet lift and a stone wall is required, extending from the dam to the lock; here it will enter the creek again, which is good boatable water, to where it becomes necessary to take out the canal. At this place, a dam of 3 feet 7 inches high is required to join an island, with another small dam to connect the island and east bank of the creek. From the dam to below the companys mill, the canal will be continued upon a level, below the mill it will be expedient to place one lock of eight feet three inches, on account of the ground falling and the steep rocky bank preventing the canal from being continued upon the same level. Above the mill, part of the canal will have to be opened by blasting of rock. From the last mentioned lock to the ferry some difficulty will be experienced, owing to the rock projecting so close to the river, it will be necessary to extend a wall down the margin of the river to near the ferry, where the river may be entered with safety by two locks of eight feet three inches lift each. It would shorten the distance of canalling in this lower end, to enter the river at the lower end of the falls at the mill, but the navigation would not be so safe and convenient as if continued to Kingsberry's ferry. (Ibid.)

Distance conversion calculated at 1 rod equals 16.5 feet and 1 link equals 0.66 feet.

161. *Southern Patriot,* March 10, 1820, 2.

162. Board of Public Works, *Plans and Progress* (1820), reprinted in Kohn, *Internal Improvement,* 75.

163. *Report of the Board of Public Works for 1821,* reprinted in ibid., 122.

164. *Report of the Board of Public Works for 1822,* reprinted in ibid., 162.

165. Blanding reported on this to the legislature: "This Canal [the Catawba Canal] was completed and went into operation last winter [winter of 1823–24], but was little used in consequence of the length of portage at Rocky Mount. This portage has never been improved by the public, in consequence of the prospect of its being much shortened, or wholly superceeded [*sic*] by the Rocky Mount Canal" (*Report of the Superintendent of Public Works for 1825,* reprinted in ibid., 344).

166. *Report of the Superintendent of Public Works for 1826,* reprinted in ibid., 466. Farrow's or Farrar's Landing was located between Mount Dearborn and Rocky Mount, on the west bank of the Catawba River.

167. Board of Public Works, *Plans and Progress* (1820), reprinted in ibid., 20.

168. Ibid., 56; *Report of the Superintendent of Public Works for 1823,* reprinted in ibid., 289.

169. Board of Public Works, *Plans and Progress* (1820), reprinted in ibid., 56.

170. Robert Leckie (ca.1775–1834), stonemason and canal contractor, was born in Perth, Scotland. In 1805 he married Mary Taylor Leckie, and they immigrated to the United States. Between 1805 and 1814 he worked in New York as a stonecutter. By 1817 he had relocated to Washington, D.C. In that year President James Monroe referred to Leckie as "the conductor of all the Arsenals U. S. and a famous Quarrier," and he was hired to quarry stone from Marble Quarry on the Potomac River in Montgomery County, Maryland, for the U.S. Capitol under Benjamin Henry Latrobe. In August 1820 Leckie accepted the job of building Landsford Canal in South Carolina. He executed this task with little trouble but had difficulties with John Couty, the assistant to Abram Blanding, over measurements for payment. At one point Leckie accused Couty of offering to overmeasure Leckie's work (and thus overpay Leckie) if Leckie gave Couty's brother, James Couty, the subcontract for the lock gates of Landsford Canal. While in South Carolina, Leckie also oversaw the construction of the York County jail and courthouse. While Leckie was building the Landsford Canal, his wife and two children died. After his work on the Landsford Canal he returned to Washington, D.C., where he worked on a number of projects. In 1829 he was appointed inspector of masonry for the Chesapeake and Ohio Canal, a position he resigned from in 1830, when it was found that the stone for the piers for the 515-foot-long Monocacy Aqueduct had deteriorated. Leckie died in 1834. See Virginia Newsome Overman, ed., biographical notes, typescript, Robert Leckie Papers, Perkins Library, Duke University Library; Allen, *History of the United States Capitol,* 119; and Kapsch and Kapsch, *The Monocacy Aqueduct.*

171. Board of Public Works, *Plans and Progress* (1820), reprinted in Kohn, *Internal Improvement,* 56.

172. *Report of the Board of Public Works for 1821,* reprinted in ibid., 122.

173. *Report of the Superintendent of Public Works for 1822,* reprinted in ibid., 162.

174. *Report of the Superintendent of Public Works for 1824,* reprinted in ibid., 345.

175. *Report of the Board of Public Works for 1823,* reprinted in ibid.,, 289.

176. Blanding reported the failed lock foundation to the legislature:

> This canal [the Landsford Canal] was finished and went into operation last autumn [1823]. The whole of the work, excepting the forebay of the second lock from the river has stood perfectly well, and was sufficiently elevated at its head to exclude the September fresh, and therefore received little injury from it. About the last of August, the forebay of the second lock gave way and caused a serious injury in the upper part of it, extending down about forty feet, and rendering it necessary to take down about one third of the masonry and to prepare a new foundation for that part of the work. I made immediate arrangements with Mr. Leckie to collect a force for re-constructing the work, which had failed; and in October attended, examined the injury and devised the plan of re-construction. It was evident that the failure had been caused by the badness of the foundation and not from any imperfection of the masonry. The head of this lock was built on a thin stratum of clay, which rested on a quick sand, out of which, issued several springs. The sand was thus gradually washed away, and left the foundation incapable of supporting the weight of the wall that rested on it. In the plan given to Mr. Leckie for re-constructing this part of the lock, I have endeavored to attain all the security which piling and timber can afford. (*Report of the Superintendent of Public Works for 1824,* reprinted in ibid., 345)

177. *Report of the Superintendent of Public Works for 1825,* reprinted in ibid., 428.

178. Letter from Abram Blanding to John Couty, November 12, 1824, South Carolina Department of Archives and History, Department of Public Works, Road, River and Canal Projects File (S208009).

179. Attachment to ibid.

180. Ibid.

181. Ibid.

182. Letter from Couty to Blanding, November 20, 1824, South Carolina Department of Archives and History, Department of Public Works, Road, River and Canal Projects File (S208009).

183. Ibid.

184. Ibid.

185. Several years later Benjamin Wright, chief engineer of the Chesapeake and Ohio Canal, wrote the following assessment of Leckie: "Mr. Leckie is a very useful man in his way—but he corrects errors with a Mall rather than a [indistinct]. Gentle means are better than harsh in managing men in this free country, and if Mr. L would correct himself in this respect and be perfectly impartial, he would do us much good" (letter from Benjamin Wright to Charles Mercer, president of the Chesapeake and Ohio Canal Company, February 24, 1830, National Archives, RG 79, entry 190).

186. In his article on this conflict, Crowson assigns the blame for it on the Board of Public Works and Couty: "Unfortunately, responsibility for the project, which was divided between himself [Leckie] and John Couty, proved to be more than Leckie had bargained for." This is not correct. Leckie was a contractor who worked under Couty's supervision, and therefore there was no division of responsibility. Crowson continues, "Couty no doubt felt the importance of his position, and attempted to use his power

as state engineer to have the project constructed in the manner preferred." Crowson adds, "Couty insisted that Leckie blast out between 3 feet and 3 feet 6 inches of rock and substitute in its place puddle (brick mortar)." Crowson is incorrect. Puddle is tamped mud, not brick mortar, and is used as a canal sealant. See Crowson, "Building the Landsford Canal." Leckie's views of his dispute with Couty are summarized in data he submitted to support his petition to the legislature:

> The following facts are submitted to shew that Couty was actuated by other motives then of a public officer in making these mismeasurements, cutting down and omissions then were his determined hostility to me which originated in a refusal by me to let his brother have an undue advantage in a contract for making the lock Gates.
>
> Couty valued the Canal when nearly done at nearly $16,000 less then the real value when finished and finding his character as a civil Engineer compromised he set to work to cut down what he had considered right in the first measurement and no doubt in my mind that the mismeasurements and omissions were made on [indistinct] to make the final measurement in 1823 correspond with his measurements and valuation of 1822, for common sense forbids us to extend to him the charity that the mistakes were made through ignorance or incapacity; If they were what are we to think of him as a civil Engineer! If willfully made what are we to [think] of him as a man:
>
> Another circumstance has a Strong bearing on this subject viz Couty declining to furnish the details of the measurement although ordered by Col. Blanding and putting me off from time to time and at last saying that he could not furnish it, the true reason. I believe to be that he withheld it on purpose fearing that if furnished it would disclose what that part of the canal unmeasured has brought to light: As a matter of right I was entitled to the details for who would have a sum of $125,000.00 decided upon by any individual without a wish to see the particulars; Gentleman, Mr. Couty ought to have furnished me with a copy of the measurement without asking it: but he was afraid that by doing so his injustice would be detected, had I been as compliant and complying as some other contractors have been I could I believe not only have had the canal accurately measured at first, but have had that liberal measurement extended to me, which Couty said he extended to others, as Mr. Culloch testified before the committee last year. ("Data for Petition, No. 2," no date, South Carolina Department of Archives and History, Department of Public Works, Road, River and Canal Projects File [S208009])

187. Crowson, "Building the Landsford Canal," 60. Crowson does not provide the date of Leckie's petition, but in 1825 the legislature ordered Blanding to remeasure Leckie's work. see *Report of the Superintendent of Public Works for 1825,* reprinted in Kohn, *Internal Improvements,* 428.

188. Footnote 6 to transcription of plaque, Landsford State Park, November 1995, Perkins Library, Duke University.

189. Robert Mills, *Statistics of South Carolina,* 158.

190. Society Hill was also called Long Bluff.

191. *Report of the Civil and Military Engineer for 1818,* reprinted in Kohn, *Internal Improvement,* A8.

192. A letter appearing in the Charleston newspapers at this time stated:

> Since August [of 1818], we have had the [Yadkin] river examined and surveyed, from Wilkes court-house, in the mountains, to the mouth of West field's Creek, above Cheraw, and find the distance two hundred and forty-seven and an half miles.—The expense of making it navigable for boats of ten tons, is estimated at $250,234, exclusive of the narrows, where for the present, we intended to make a turnpike road; distance about 7 miles; expence about 2000 per mile. $63,400 of the above sum is required to be expended between Sneedsborough and Cheraw Hill; distance 6 ½ miles, the whole distance, except 200 yards in South Carolina. The amount of our stock subscribed, is about $170,000. We again opened books, which will continue open until the 1st Thursday in January. We hope, in that time, to make up $250,000. ("Extract of a letter from one of the gentlemen, composing the above Company," October 26, 1818, *Southern Patriot,* November 5, 1818, 3).

193. Ibid. A portion of this navigation was to be a seven mile long turnpike at "The Narrows" to be constructed at a cost of $14,000. At the time the Yadkin Company was seeking a charter from the South Carolina legislature to extend their charter to Georgetown where steamboat service would connect the North Carolina trade with Charleston. By 1819 the Yadkin Navigation Company had contracted with John Hixon and Hiram Jennings to undertake this work to the South Carolina / North Carolina state line. See *Southern Patriot,* February 15, 1819, 2. Little of this work was completed. Weaver reports that the State of North Carolina purchased $25,000 of the stock of the company, but by 1828 the North Carolina state legislature had passed a resolution instructing the North Carolina Board of Public Works to investigate what happened to the $25,000. The company was again investigated in 1833. There were additional efforts to open the Yadkin River to navigation. The North Carolina superintendent of public works reported that more money had been spent on this navigation than any other in North Carolina except the Roanoke and Cape Fear rivers and without benefit to the state. See Weaver, *Internal Improvements in North Carolina,* 65–66.

194. The newspaper extract read:

> Gen. Pearson (one of our directors) will attend your Legislature. The object in view, is to obtain from South-Carolina an extension of our charter, so as to take in the whole river to Georgetown. If, under their present plan of internal improvement, that is not admissible, then have the exclusive right to such improvement as they may make by canals, leaving the bed of the river to be improved by the State. The distance from Cheraw to Georgetown by land, is 103 miles, by water, it is called 270 miles, at least one third of that distance can be saved by canals at

> a less expense, than clearing the river of logs. If we cannot obtain that, (and if the State will do it we do not want it) we hope to have granted to us, the right of cutting a canal and looking down from Sneedsborough to Cheraw Hill, a distance of 6½ miles. ("Extract of a letter from one of the gentlemen, composing the above Company," October 26, 1818, *Southern Patriot,* November 5, 1818, 3)

195. *Report of the Civil and Military Engineer for 1819,* reprinted in Kohn, *Internal Improvement,* 6.

196. Ibid.

197. The board reported to the state legislature, "Learning, however, that, notwithstanding the efforts of the North Carolina Yadkin Company, the navigation above the line will, in all probability, remain closed for some time." Board of Public Works, *Plans and Progress* (1820), reprinted in ibid., 20. In fact the company only built a few miles of canal along the Yadkin River. In 1823 the company became embroiled over allegations that the stagnant water in the Yadkin Canal was causing disease and death in Sneedsboro, North Carolina:

> This little village, about 14 miles above Cheraw on the Pee Dee, has been unusually sickly this season, which is conceived by the faculty, to have originated from the stagnant water in the Yadkin canal, the outlet of which, to the river, is through, or near the compactly settled part of the town; great oversight as demonstrated in this almost exterminating disease, is attributable to the commissioners who should, in the early part of the summer, have ordered the lower sluices to be opened, by which a constant current, and that of fresh and wholesome water, would have been kept up, and would probably have prevented the serious and alarming consequences which *have* resulted and continued to result to the population of a district five or six miles in extent. The family of the late Mr. Knox have been peculiarly unfortunate. Mr. Knox and Mrs. Knox, their eldest daughter, a young lady of about 23 years of age, and child of 7 or 8 years, all have died of fever, within a few days of each other; many other families have suffered severely, and since the death of Miss Johnson, daughter of Wm. Johnson. Esq. most of the population have removed farther up the country. (*Southern Patriot,* October 14, 1823, 2)

198. Since its incorporation in 1820, the official name of the town was Cheraw. However, it was frequently referred to by its earlier name, Chatham, or by the name Cheraw Hill.

199. Board of Public Works, *Plans and Progress* (1820), reprinted in Kohn, *Internal Improvement,* 20.

200. "General Williams continues the operations on this river, with a force, which, it is expected, will finish it this year" (*Southern Patriot,* March 10, 1820, 2).

201. Blanding's instruction to General Williams for work on the Great Pee Dee stated:

> Six Boats are to be put into complete order, and furnished with the necessary implements and machinery for raising & removing Logs from the bed of the

river. One Superintendent and two Captains are to be engaged to conduct the work if so many be necessary, and each Boat to be manned with ten able-bodied Negroes—There shall be one black Coxswain to each Boat who shall receive two Dollars per months over & above his Wages, whenever in the opinion of Genl. Williams he shall by his good conduct have merited this gratuity. Due economy will be used in the purchase and distribution of the provisions, but the quality and quantity of the rations are left to the discretion of Genl. Williams. The Wages to be given to the Men are left likewise to his discretion, as he will be the best judge of their qualifications, and of the value of their labour. A regular Pay Roll will be kept, and together with a statement of the expenditure in Tools and provisions will be forwarded monthly to the Office of the Board of Public Works at Columbia. Due attention must be paid to the sick, and deductions for sickness shall be made only when more than one week's labor is lost in the month. The Work is to be commenced at the lowest point of the River where obstructions exists, and proceed upwards. The width of the Channel to be cleared ought not to exceed one hundred and fifty feet. In removing Logs that will float, they are either to be hauled on Shore and placed above high water mark, or cut into short pieces & suffered to float down the River, care being taken to prevent their forming rafts of drift wood. Those Logs which are too water sodden to float must be deposited in Eddies or Old Rivers where practicable but always out of the course of Boats ascending the River. All the Trees on the banks of the River that hang over the Channel, or are from their position likely soon to fall into the River, must be cut down and disposed of, and such as are exposed to be undermined, and are so situated as to be in danger of falling into the River at any future time must be girdled. This work must be particularly attended to when the State of the River will not admit of working in its channel. Whenever any unforeseen obstacle shall arise, General Williams will report the same to the Board, and he will in no case attempt to remove Sand Bars, Stone ledges, or cut through a neck of land without the advice and concurrence of the Board. In carrying on the work Genl. Williams will take care to cause as little injury as possible to the Inhabitants on the River and he will stipulate with the persons from whom he hires Slaves, that in case of their committing any outrage upon the property of the Inhabitants, he shall be at liberty to discharge them. ("Instructions for Gen. Williams in Conducting the Work on the Pee Dee," February 17, 1820, attached to "Instructions for Col. Wm. Nixon in Conducting the Work on the Wateree River," South Carolina Department of Archives and History, Department of Public Works, Miscellaneous Papers, February 17, 1820 [S208006])

202. These workmen were slaves hired for the purpose. Williams reported, "Unfortunately, one person fell overboard, while at work, and was drowned. The owner is in very moderate circumstances, and his loss is respectfully submitted for your consideration" (letter from Williams to J. R. Poinsett, November 3, 1820, as published in Board of Public Works, *Plans and Progress* [1820], reprinted in Kohn, *Internal Improvement*, 58).

203. The Charleston newspapers reported: "It will no doubt be highly satisfactory to those who feel interested in the opening and improving of our Internal Navigation, to know that the *Pee Dee River,* under the care and immediate superintendence of that very active and truly zealous citizen, David R. Williams, has been entirely cleared, and all the Workmen discharged. This work has opened a very profitable interchange between Chatham, Georgetown and Charleston, by which much of the produce of North-Carolina finds an advantageous market" (*Southern Patriot,* August 21, 1820, 2).

204. *Report of the Superintendent of Public Works for 1823,* reprinted in Kohn, *Internal Improvement,* 277.

205. The board reported:

> An examination of the Peedee, made as early as the state of the water would permit, presents this stream in a very favorable point of view. Its importance to the trade of South-Carolina must be evident when we regard its connection with the extensive back country, which it waters both in this state and in North-Carolina. It extends with a depth of four feet water in a common river to Cheraw, one hundred and twenty miles from the seaboard. The circuitous course of the stream nearly doubles this distance, yet, the Steam Boat descends from Cheraw to Georgetown, 14 miles from the sea, in three days, and returns in five. This is done with ease, when there is a small swell, but when the river becomes low, there are several points where sand or gravel bars reduce the depth of water, and obstruct the navigation of boats of the largest class. The worst of these obstructions is at Thompson's creek, about two miles below Cheraw. Measures have been taken to deepen the channel at this place, and it is the intention of the board to continue the work when the state of the river will permit, until four feet water is procured from Cheraw to Georgetown. (*Report of the Board of Public Works for 1821,* reprinted in ibid., 121; extract from "Report of the Board of Public Works," *Southern Patriot,* December 10, 1821, 2)

206. Blanding reported: "It [the Great Pee Dee River] has a steam and team boat navigation to its first fall, about half a mile above the town of Cheraw" (*Report of the Superintendent of Public Works for 1823,* reprinted in ibid., 277). "Team boats" are not defined but were probably horse-powered boats.

207. These fifteen obstacles in the Great Pee Dee River were:

1. Thompson's Shoals, two and a half miles below Cheraw and opposite Thompson's Creek, entering the Pee Dee River on its right bank—a number of obstructions, primarily sand bars.
2. Thomas N. Ellerbe's Shoal, two and a half miles below Thompson's Creek—a sandbar.
3. Board Landing or Griffin's Shoal, two miles below Ellerbe's. Considered one of the greatest difficulties in navigating the Pee Dee, it consisted of a whirlpool and obstructions caused by trees falling into the river as the right bank was eroded.
4. Zachariah Ellerbe's Shoal, three hundred yards below the end of Board Landing. It consisted of a sandbar in the middle of the river. As at Board Landing, large trees were continually undermined and thrown into the channel.

5. Terrel's Shoals, half a mile down river from Zachariah Ellerbe's Shoal. A bar made a crooked, narrow and difficult channel.
6. Thomas's Fishing Bar, two miles below Terrel's Shoals. A bar extended from the right bank almost across the entire river. Below this is what was called Cheraw Bay, seven or eight miles of good river to Society Hill (also called Long Bluff).
7. Whirligig, four miles below Society Hill at the mouth of Buckeler's Creek. Here the river makes a sudden turn to the right with a bar on the left. At the mouth of Buckeler's Creek was a rocky bar or shoal.
8. Edward's Shoal, about one and a half miles below the Whirligig. Here there was a bar on the right bank and at its lower end another on the left bank.
9. Browker's Falls or Shoal, located one mile below Edward's Shoal and similar in nature.
10. Roger's Shoal, located one mile below Browker's Falls. A bar extended from the left shore.
11. Peterkin's Bar, located two miles below Roger's Shoal. The bar was in the middle of the river.
12. Campbell's Shoal and Fishing Place was located half a mile below Peterkin's Bar. Here there were opposing sand and gravel bars from both banks. At their lowest end there was a third bar, in the middle of the river.
13. Pounoy Shoal was a half mile below Campbell's Shoal. Shoals extend from both the right and left banks.
14. Sugar Loaf Shoal, located seven miles below Pounoy Shoal. Here a bar extended from the right bank.
15. The Pocket, fourteen miles below Sugar Loaf Shoal, was the last obstacle in navigation between Cheraw and Georgetown. Three bars made the channel crooked and difficult to traverse. (Ibid., 278–79)

208. *Report of the Superintendent of Public Works for 1824,* reprinted in Kohn, *Internal Improvement,* 342.

209. Ibid., 343.

210. *Report of the Superintendent of Public Works for 1828,* reprinted in ibid., 565.

211. Ibid.

212. *Report of the Superintendent of Public Works for 1823,* reprinted in ibid., 285.

213. Letter from General David R. Williams to J. R. Poinsett, November 3, 1820, as published in the Board of Public Works, *Plans and Progress* (1820), reprinted in ibid., 58. Work undertaken on the Waccamaw River, Black River, and Black Mingo was described as follows:

Waccamaw River: "Waccamaw river is navigable from Winyaw bay to the North-Carolina line, being between seventy and eighty miles. The only obstruction to its navigation are logs, which have been suffered to accumulate in the river, and one sand bar extending up the stream nearly four hundred yards, and beginning about 50 miles from the mouth of the river. There is never less than five feet water in any part of the Waccamaw" (*Report of the Civil and Military Engineer for 1818,* reprinted in ibid., A14). "In April last [1826] I again visited it [the Waccamaw River], and formed a contract with Alexander W. McRa, for continuing the boat navigation to the North Carolina line. I procured this work to be done within the appropriation, leaving a balance of $265. The result of these contracts when completed, will be that there will be a schooner navigation from Georgetown to Porter landing [fourteen miles above Conway], a distance of 75 miles by water, and a boat navigation above Porter's landing to the North Carolina line, a distance of more than 35 miles by water. This work will have to be examined and received next year [1827], when it will be finished" (*Report of the Superintendent of Public Works for 1826,* reprinted in ibid., 450).

Black River: "Black river is navigable for boats carrying forty or fifty barrels of Rice, as far as the village of Kingstree, forty-five miles in a direct line, from Winyaw Bay; and for smaller craft to Lowry's bridge, twelve miles above Kingstree. This river might be opened twenty-five miles higher up, as the only obstructions to its navigation, are logs and sand bars" (*Report of the Civil and Military Engineer for 1818,* reprinted in ibid., A14). "In my last years report, I stated that the probable cost of clearing this river [the Black River] to Lowry's bridge, would be $3000. I attended at Kingstree, and on the 25th April last [1826], made a contract for this work, at $1,488, with Col. William Salters. The contract stipulates for removing all obstructions up to Lowry's bridge, on the lower line of Sumter District, except the rocky bars near the lower bridge at Kingstree" (*Report of the Superintendent of Public Works for 1826,* reprinted in ibid., 462).

Black Mingo Creek: "Black Mingo creek enters Black river 15 or 16 miles above Georgetown. It was formerly navigated by schooners to Wiltown old bridge, 7 miles by land and 17 or 18 miles by water from its mouth. Wiltown is now deserted and the bridge removed one mile lower down the creek. Here the schooner navigation must stop, but boats carrying 150 bales of cotton might ascend three miles higher. But to effect this, the present obstructions must be removed. These consist of logs alone; they are so numerous that a schooner cannot ascend or descend, nor has any done so for the last 25 years. This obstruction to the navigation was one of the causes of the decline of Wiltown. The extent of the country to be benefited by this navigation is small and it has no commerce. There is, however, much valuable cypress timber in this swamp, which can be brought to market only down this creek. The navigation might be extended to Indiantown 12 miles above Wiltown, was there an adequate object for it. But most of the pine timber was destroyed by the gale of 1822, and this timber furnished the only articles, which require a boat navigation" (*Report of the Superintendent of Public Works for 1825,* reprinted in ibid., 419).

There are no subsequent reports that would indicate that Blanding undertook improvements on Black Mingo Creek.

214. Board of Public Works, *Plans and Progress* (1820), reprinted in ibid., 72. Work undertaken on Black Creek, and Lynches River (known then as Lynch's (or Lynche's) Creek:

Black Creek: "This creek can be rendered navigable by boat 60 feet long, 12 wide and drawing 2 feet water in the driest seasons, for 35 miles by land from its confluence with the Pee Dee. Its only obstructions are logs and sand bars; As soon as Gen. Williams completed the navigation of Pee Dee, one set of flats and one boat were placed on this creek, under Rasha Cannon, esq. He commenced work on the 7th September last [1820], and has ascended 15 miles by water, removing in this distance, all the obstructions and clearing the banks of all the trees likely to fall in, or obstruct the passage of boats" (Board of Public Works, *Plans and Progress* [1820], reprinted in ibid., 61). "Black creek from Darlington court-house down to its junction with Peedee was put under contract in May last at public auction. The appropriation of $1000 was sufficient for the purpose. The work has been carried on during the summer with advantage, but is not entirely done" (*Report of the Superintendent of Public Works for 1827,* reprinted in ibid., 514).

Lynches River: "Lynche's Creek, which falls into the Great Pedee, about fifty-two miles from Winyaw bay, is obstructed near its mouth by logs and drift-wood, which have been thrown up by the eddy water of the Pedee river in time of freshets. These obstructions extend nearly three miles up the creek. The navigation might be opened with little labor, for boats drawings three or four feet, to the confluence of Big and Little Lynches Creek, a distance of eighty-two miles by land. Each of these branches might be rendered navigable ten or twelve miles above the confluence" (*Report of the Civil and Military Engineer for 1818,* reprinted in ibid., A9).

Blanding explained his decision not to open the lower portion of Lynches River: "At the mouth of the creek it is greatly obstructed by immense rafts, which are thrown into it from the Pee Dee, which often has great rises of water in it, when there is no corresponding rise in the creek. Were these obstructions removed, they would again accumulate, and in a short time the same expense must be incurred in their removal. About eight miles above the mouth of the creek, there is a low swamp leading from it, which, in 350 yards, reaches a succession of small lakes that empty into Clark's Creek, and that into Pee Dee river, two miles above Britton's ferry. Were the channel opened in this direction, it would be about eight miles shorter than by its present mouth, and would not be subject to be choaked up by the rafts from Pee Dee" (*Report of the Superintendent of Public Works for 1825,* reprinted in ibid., 416).

215. Blanding reported on the work on the tributaries: "The works on Lynch's creek, Black river and Waccamaw were not finished, and could not be received, and since the close of the sickley season I have not had time again to examine them, altho' I have received notice from some of the contractors that their work was ready for examination" (*Report of the Superintendent of Public Works for 1827,* reprinted in ibid., 514).

216. Letter to the editor signed "A South Carolinian," *Southern Patriot,* February 11, 1819, 3. The Winyaw and Wando Company was the second company chartered to build a canal between the vicinity of Georgetown and Winyah Bay to the Santee River and from there to the Cooper or Wando River thus opening an inland passage between Winyah Bay and Charleston. The first was the "Company for the Inland Navigation, from Sampit into Santee, and from Santee into Cooper or Wando River," chartered by the state legislature on January 9, 1810. See Sampit and St. James Santee Canal Company, *Rules and Regulations;* McCord, ed., *The Statutes at Large of South Carolina,* 7:580–82.

217. *Southern Patriot,* May 21, 1819, 3.

218. The proposed passage from the North Santee to the Wando River to Charleston was (proceeding south to north):

> The first canal delineated on the plan (A. B.) commences at the head of navigation on Wando river, about twenty-four miles from Charleston by the windings of the river, and passes through the swamp, which is formerly the head waters of Wando River and Owendaw Creek. The canal enters Owendaw at the head of boat navigation. Following the Owendaw to within a quarter of a mile of Bull's bay, it takes a northeast course through the creeks C. D. E. F. G. H. I. until it enters Aligator Creek at (I.) From the point (L.) which is nearly opposite the entrance of Alligator Creek, a canal of twenty chains in length, will connect South Santee with Atchinson's creek, a bold stream falling into North Santee at (M.). Crossing the North Santee, the navigation passes through Duck creek (N.) It is then proposed to ascend Minim creek as far as the first bend, (O.) and from thence take a Northeastwardly course, until it enters Winyaw bay at (Q.) six or seven miles below Georgetown. (*Report of the Civil and Military Engineer for 1818,* reprinted in Kohn, *Internal Improvement,* A17)

The map to which the above text refers has not been found and is believed to be non-extant.

219. *Report of the Superintendent of Public Works for 1823,* reprinted in ibid., 271–72.

220. Ibid., 272.

221. *Report of the Civil and Military Engineer for 1818,* reprinted in ibid., A18. Initially brick locks were planned.

222. Board of Public Works, *Plans and Progress* (1820), reprinted in ibid., 21. The report stated: "Begun in 1819, and the work has since been carried on steadily. From the nature of the ground a swamp of large cypress, and from the difficulty of hiring laborers, the cutting has advanced slowly, and not quite two miles are completed."

223. *Report of the Superintendent of Public Works for 1823,* reprinted in ibid., 271. In addition the state paid $19,529.81 for the island passage from the Santee River to Charleston.

224. Ibid., 270. Blanding described the route: "The boats, rafts, etc, which descend the *North, South* and *Main Edisto's* now reach Charleston by a circuitous route, which carries then along Dawho river, to North Edisto inlet, and Wadmalaw to Stono river, and thence by Wappon Cut to Charleston. A part of this passage is exposed, and dangerous for small boats. Rafts have been frequently lost here" (*Report of the Superintendent of Public Works for 1825,* reprinted in ibid., 374–75).

225. *Report of the Superintendent of Public Works for 1823,* reprinted in ibid., 269–70.

226. Ibid., 270.

227. Ibid.

228. Ibid.

229. *Report of the Superintendent of Public Works for 1825,* reprinted in ibid., 369.

230. Ibid., 372.

231. Ibid., 374. Blanding provided a breakdown of these costs:

> The total cost of this canal [which would provide water to Charleston and navigation] may therefore be thus stated:

$4\frac{1}{4}$ miles of canal at $7,500 per mile,	$31,875
23 chains [1518 ft.] of ditto at the ridge,	8,000
27 miles of canal at $3,475	93,825
40 chain [2640 ft.] of embankment, at	16,000
2 miles of Do.	30,500
Basin at Charleston lines,	7,500
Guard lock of freestone,	12,000
Aqueduct, stone piers and cypress superstructure,	$13,800
	$213,500
The places which will require culverts, and waste wiers cannot be determined till the plan of execution is fixed: they may amount to 20 of each; the first will cost $400 and the latter $200 each, or	8,000
	$221,500

. . .

But if this work should not embrace the supply of Charleston with water, the canal will then enter the Ashly [*sic*] river, and the cost may then be thus stated:

4 miles, 20 chains [1320 ft. or 4.25 miles] of canal, at 7,500 per mile,	$31,875
23 chains [1518 ft.], do	8,000
10 miles do. at $3,474 per mile,	34,750
1 mile, 40 chains [2640 ft.] embankment,	22,500
1 guard and 2 lifting locks, at $12,000 each,	36,000
For culverts and waste wiers,	6,000
	$139,125
If masonry is of brick it will reduce the expense,	12,000
The work would then cost	$127,125

232. Robert Mills, *Statistics of South Carolina,* 159. Wilson provided a more detailed description of the navigation possibilities of the lower Edisto River:

The Edisto river empties itself into the ocean by two mouths, The North Edisto discharges its water between Simmons's Island and Clarks Island, and South Edisto enters the ocean between Eddings's Island and Fenwick's or Tucker's Island. The Wadmelaw river branches from the North Edisto about eight miles from its mouth. The inland navigation passes through this branch into Stono river, which takes its rise in the swamps and marshes between the Edisto and Ashley, and communicates with Charleston harbor by Wappoo Creek.

There are no obstructions to this navigation up the branches of the Edisto about fifteen mile[s] from the ocean and vessels drawing five or six feet water, trade to Jacksonborough, ten miles higher up.

From this point, the navigation continues good to Parker's Ferry, about forty miles from the ocean, where the influence of the tide ceases. Above Parker's ferry, the obstructions commence, and consist of sand bars and logs; but during eight

> months in the year, the navigation is good for rafts and boats to the confluence of the North and South Branches, about eighty miles in a direct line from the ocean. (*Report of the Civil and Military Engineer for 1818,* reprinted in Kohn, *Internal Improvement,* A14–A15)

233. *Report of the Civil and Military Engineer for 1819,* reprinted in ibid., 13.

234. *Report of the Civil and Military Engineer for 1818,* reprinted in ibid., A15.

235. Ibid.

236. The instructions Blanding gave to James Head for clearing the South Edisto River were very similar to those he gave to Colonel W. J. Myddleton for clearing the Congaree and Santee rivers except that in Head's instructions, Blanding specified that, "each set of Flatts [should have] Ten Slaves." (Instead of the fifteen slaves specified in Colonel Nixon's instructions). A copy of Head's instructions was sent to David Williamson on the North Edisto River, and the same instructions were also applied to the work undertaken to clear the North Edisto. See "Instructions for Mr. James Head, in conducting the Works on the South Edisto River," South Carolina Department of Archives and History, Department of Public Works, Road, River and Canal Projects File, Miscellaneous Papers, February 15, 1820 (S208006).

237. Board of Public Works, *Plans and Progress* (1820), reprinted in Kohn, *Internal Improvement,* 21.

238. Robert Mills, *Statistics of South Carolina,* 156–57.

239. Ibid., 157.

240. "Communications," *Southern Patriot,* April 19, 1819, 2.

241. Board of Public Works, *Plans and Progress* (1820), reprinted in Kohn, *Internal Improvement,* 21.

242. Wilson described this channel in his 1818 report to the state legislature:

> By the second article of the convention between these states, the navigation of the river Savannah, at and from the bar and mouth along the Northeast side of Cock-Spur Island, and up the direct course of the main northern channel, along the northern side of Hutchinson's Island, opposite to the town of Savannah, to the upper end of the said Island, and from thence up the bed or principal stream of the said river, to the confluence of the rivers Tugoloe and Keowee, and from the confluence, up the channel of the most northern stream of the Tugoloe river to its source, and back again by the same again by the same channel to the Atlantic Ocean, is hereby declared to be hereafter equally free to the citizens of both states, and exempt from all duties, tolls, hindrance, interruption or molestation whatsoever, attempted to be enforced by one state on the citizens of the other; and the rest of the river Savannah, to the southward of the foregoing description, is acknowledged to be the exclusive right of the state of Georgia. (*Report of the Civil and Military Engineer for 1818,* reprinted in Kohn, *Internal Improvement,* A13)

243. Board of Public Works, *Plans and Progress* (1820), reprinted in ibid., 21.

244. Ibid.

245. *Report of the Civil and Military Engineer for 1818,* reprinted in ibid., A13.

246. Ibid.

247. Board of Public Works, *Plans and Progress* (1820), reprinted in ibid., 21.

248. The *Southern Patriot* reported: "We learn that the President of the Board of Public Works met the Commissioners on the part of Georgia last week, and that arrangements were made for the commencing the work between Petersburgh and Augusta immediately" (*Southern Patriot,* July 7, 1820, 2).

249. *Report of the Superintendent of Public Works for 1823,* reprinted in Kohn, *Internal Improvement,* 285.

250. Phillips, *A History of Transportation in the Eastern Cotton Belt,* 93–94. Roads, however, continued to improve and became a viable alternative to canal freight.

251. Board of Public Works, *Plans and Progress* (1820), reprinted in Kohn, *Internal Improvement,* 76–77.

252. *Southern Patriot,* November 25, 1820, 2.

253. Board of Public Works, *Plans and Progress* (1820), reprinted in Kohn, *Internal Improvement,* 77.

254. "Extract of a Letter from the Commissioner of Public Works [Abram Blanding]," dated July 29, 1820, quoted in *Southern Patriot,* August 11, 1820, 2.

255. Board of Public Works, *Plans and Progress* (1820), reprinted in Kohn, *Internal Improvement,* 49.

CHAPTER 5. *Operating the South Carolina Canal and River-Navigation System, 1819–1828*

1. "*COLUMBIA*—For Sale," *Southern Patriot,* October 11, 1823, 3.

2. Advertisement for the sale of the steamboat *Columbia,* ibid., March 8, 1821, 3. By 1822 the steamboat *Columbia* was on the Edisto, Beaufort, and Savannah run, "having undergone a th[o]rough repair, with considerable alterations" (ibid., February 23, 1822).

3. "Auctions. Steam Boat *Columbia,*" ibid., October 20, 1826, 3.

4. Towboats were used to carry cargo to and from Granby, immediately below Columbia, and elsewhere. Typically a steamboat would have four towboats. Two of them would be left at Granby (or Charleston) and the steamboat would pick up two other towboats, thereby allowing the towboats left to be unloaded and reloaded at a leisurely pace without delaying the departure of the steamboat.

5. Letter to the editor, dated June 30, 1819, *Southern Patriot,* July 3, 1819, 3.

6. "Passage for Edisto, Beaufort, and Savannah," ibid., February 23, 1822, 3.

7. "*COLUMBIA*—For Sale," ibid., October 11, 1823, 3.

8. "Steam Boat *Columbia,*" ibid., June 22, 1824, 3.

9. Ibid.

10. The newspaper account of the grounding and sinking of the *Columbia* reads:

> The Steam Boat Columbia, captain Clark, with one Tow Boat, laden with Merchandise for Cheraw, left the wharf at 5 o'clock yesterday afternoon [October 21, 1824], for Georgetown and Cheraw. She proceeded by the Inland Passage to within two miles from Sullivan's Island, and there anchored both head and stern; during the early part of the night it was discovered that she had grounded, and had taken a considerable heel to the larboard side, and gradually slid off into deep water, which rushed into her ports when she soon filled and sunk, it then being about 12 o'clock at night. Mrs. Rantin and daughter, (about 9 years of age)

wife of Mr. William Rantin, formerly of this city, and at present a resident of Georgetown; a Wench and child, belonging to Mr. John Macdonald of Cheraw, who were in the after Cabin, were unfortunately drowned. Mr. Flagg and Doctor Hume of this city, also passengers, as well as the captain and all the crew, with much difficulty saved themselves in the Tow Boat. The Columbia had a trifling value of cargo in her cabin, which, together with most of the baggage of the passengers and crew, remains on board, but in all probability will be saved, as hopes are entertained that the Boat may be raised. Captain Clark, states that he had a full watch on deck, but from the sudden filling of the Columbia, it was even with difficulty that their lives were saved. The Columbia we understand is insured in New-York, and is owned by Messrs. Beers & Buanell, of this city. We learn from one of the passengers, that on awaking, he found the water waist deep in the cabin, and that before leaving it, an effort was made with the other passengers to save the lives of the females. ("Melancholy Occurrence," *Southern Patriot,* October 22, 1824, 2)

11. "For Cheraw, SC via Georgetown, The Steam Boat Columbia," ibid., March 27, 1826, 3; "Auctions, Steam Boat Columbia," ibid., October 20, 1826, 3.

12. "Steam Boat *South Carolina* [For Sale]," ibid., September 15, 1823, 3.

13. "Fire!" ibid., June 30, 1821, 2.

14. The steamboat was not insured, and on July 10, 1821, the stockholders of the Congaree and Santee Steam Boat Company voted to auction it off. The auction was held on August 13, 1821. See "Congaree and Santee Steam Boat Company," ibid., July 30, 1821, 3.

15. *Southern Patriot,* April 8, 1822, 2. By 1822 a third boat was on the Granby-Charleston run, the steamboat *Patent.* See "Freight for Granby, and Columbia," ibid., April 6, 1822, 3.

16. On June 23, 1820, the *Southern Patriot* reported that the *South Carolina* had made a voyage from Charleston to Columbia in ten and a half days, noting that "she was detained three days in the river getting fuel." *Southern Patriot,* June 23, 1820, 2.

17. *Report of the Superintendent of Public Works for 1823,* reprinted in Kohn, *Internal Improvement,* 266.

18. "Steam Boat *South Carolina,*" *Southern Patriot,* September 15, 1823, 3.

19. Ibid.

20. "For Hamburg and Augusta, The Steam Boat *South-Carolina,*" ibid., March 15, 1826, 3.

21. "For Cheraw via Georgetown, The Steam Boat *South-Carolina,*" ibid., October 19, 1826, 3.

22. "Steam Boat *Maid of Orleans,*" ibid., October 28, 1822, 3.

23. "*Maid of Orleans,*" ibid., March 5, 1823, 3.

24. "For Hamburg and Augusta, The Steam Boat *Maid of Orleans,*" ibid., October 28, 1824, 3.

25. *Southern Patriot,* March 16, 1820, 3.

26. Ibid., January 22, 1820, 3.

27. Ibid., January 14, 1820, 3. Until it was sold in 1820, the *Charleston* was operated by the Charleston Steam Boat Company.

28. "For Savannah via Beaufort," ibid., January 3, 1821, 3. Edisto, as a steamboat destination, was added in 1822. See "The Steam Boat *Charleston,*" ibid., February 18, 1822, 3.

29. "For Granby, The Steam Boat *Charleston,*" ibid., February 11, 1823, 3.

30. "Freight for Granby and Columbia, The Steam Boat *Patent,*" ibid., April 6, 1822, 3.

31. According to the newspaper account, "the steam-boat *Charleston,* on her way from Georgetown to Cheraw, run one of her Tow Boats upon a snag: when she immediately sunk, laden with valuable merchandize in that place. Loss variously estimated, from two to five thousand dollars" (*Southern Patriot,* November 18, 1823, 2).

32. "For Sale, Two BOATS," ibid., April 18, 1826, 3.

33. "Auctions. Steam Boat *Charleston,*" ibid., August 23, 1826, 3.

34. "The Steam Boat *Charleston,* owned by Capt. Bonnell, was launched this morning from Marsh's wharf. She went off in handsome style" (ibid., September 16, 1826, 2).

35. "The Steam-Boat *Hamburg,*" ibid., March 27, 1823, 3; "For Hamburg, S.C.," ibid., December 31, 1823, 3.

36. "Extract of a Letter from Charles Clarke, Engineer of the Steam Boat *Hamburg,*" ibid., June 5, 1826, 2.

37. "For Augusta and Hamburg, the Steam Boat *Hamburgh,*" ibid., October 13, 1826, 3.

38. "For Hamburg, The New Steam Boat *Wm. Lownes,*" ibid., October 28, 1823, 3; "For Hamburg, via Beaufort, and Savannah, The Steam Boat *Commerce,*" ibid., May 16, 1823, 3.

39. "Auctions, The Steam Boat *Commerce,*" ibid., March 14, 1826, 3.

40. "For Hamburg and Augusta, The New Steam Boat *Henry Schultz,*" ibid., September 17, 1824, 3.

41. "Valuable Steam Engine at Auction," ibid., May 4, 1825, 3.

42. "On account of the Underwriters, Public Sale at Augusta," ibid., April 28, 1825, 3; "Valuable Steam Engine at Auction," ibid., May 4, 1825, 3.

43. "Notice," ibid., September 14, 1825, 3.

44. "For Augusta and Hamburg, The Steam Boat *General Greene,*" ibid., April 27, 1826, 3; "For Augusta and Hamburg, The Steam Boat *Macon,*" ibid., April 13, 1826, 3.

45. "Congaree and Santee Steam Boat Company," ibid., January 21, 1819, 2. Joseph Johnson was chairman.

46. "Congaree and Santee Steam Boat Company," ibid., July 30, 1821, 3.

47. "Freight for Granby," ibid., January 12, 1822, 3.

48. "For Columbia, The Steam Boat *Spartan,*" ibid., May 9, 1825, 3.

49. "We understand that the steam boat *Pee Dee* has burst her boiler" (ibid., January 16, 1826, 2).

50. "There were no lives lost by the bursting of the boiler of the Steam Boat *Pee Dee*" (ibid., January 26, 1826, 2).

51. "Auctions, Steam Boat *Pee Dee,*" ibid., March 17, 1826, 3.

52. "For Augusta and Hamburg, the Steam Boat *Pee Dee,*" ibid., April 3, 1826, 3.

53. "Pee Dee Steam Boat Company," ibid., March 16, 1821, 1.

54. "The Cheraw Spectator Says," ibid., October 3, 1826, 2. The steamboat *North Carolina* was built in Fayetteville, North Carolina, in 1819. It displaced 115 tons. Its first home port was Georgetown, South Carolina. It was sunk in late 1823, later raised, and operated from 1824 to 1829, when it was abandoned. See "NC Business History—North Carolina Steamboats 1812–1839," at http://www.historync.org/NCsteamboat list1812 - 1849.htm (accessed May 14, 2009).

55. "Steam Boat Navigation," *Southern Patriot,* March 17, 1819, 2.

56. Ibid., April 8, 1820, 2.

57. "Steam Boat," ibid., February 25, 1823, 2.

58. "Launch," ibid., September 16, 1828, 2.

59. "A Steam Boat to be called the *General Jackson,*" ibid., October 13, 1828, 2.

60. "For Sale, To Close the Steam Boat Concerns," ibid., December 22, 1828, 3. The steamboat *Augusta* had a long life. In 1838 she was reported still operating. The *Augusta* displaced 206 tons and was 113 feet long with a breadth of 29 feet and a depth of 7 feet and 7 inches. See "Letter from the Secretary of the Treasury, Transmitting in Obedience to a Resolution of the House of the 29th of June last, information in relation to Steam-Engines, etc.," United States Serial Set No. 345, 25th Cong., 3d sess., Exec. Doc. No. 21 (December 13, 1838), 266: "Return Respecting Steam Engines Used in Steamboats in the District of Charleston," on-line at http://www.maritimehistory ofthegreatlakes.ca/Bib.25p?PubID=13 (accessed July 3, 2009).

61. Board of Public Works, *Plans and Progress* (1820), reprinted in Kohn, *Internal Improvement,* 73.

62. *Report of the Board of Public Works for 1822,* reprinted in ibid., 168–69.

63. "For Savannah via Beaufort, The Steam Boat *Charleston,*" *Southern Patriot,* January 3, 1821, 3.

64. *Report of the Superintendent of Public Works for 1823,* reprinted in Kohn, *Internal Improvement,* 272–73.

65. A report from Columbia dated October 4 noted: "A considerable fall of rain on Wednesday last, has raised the river a little, but we fear nor sufficient to open the intercourse between this place and Charleston for boats, for any length of time, if at all. The river has for some time past been uncommonly low; for notwithstanding the rains have been sufficient to make a good crop, they have been too light to produce much effect on the water course—Gaz." (*Southern Patriot* for October 7, 1828, 2).

66. "Alabama Steam Boats," *Southern Patriot,* July 9, 1828, 2.

67. "New Pettiauger to Be Sold Cheap," *City Gazette or Daily Advertiser* (Charleston, S.C.), January 21, 1797, 3; "For Sale, An Excellent Congaree-Built Boat," ibid., March 6, 1800, 2; "For Sale, A First Rate New Schooner Built Congaree Boat," *Southern Patriot,* March 12, 1818, 2; "For Sale A Congaree Boat," *City Gazette or Daily Advertiser,* August 2, 1791, 3.

68. "For Sale, A Good Size Batteau," *Southern Patriot,* February 20, 1821; "For Sale, A Strong and New Bateau," ibid., June 28, 1823, 3; "Batteau Boat Stolen." *City Gazette,* November 7, 1792, 4.

69. "A Canoe," *City Gazette,* March 7, 1800, 3; "Canoe Stolen." *Southern Patriot,* November 20, 1823, 3.

70. [Canal Boats Arriving at Charleston], *Southern Patriot,* November 25, 1822, 3; "For Sale or to be Hired, A Canal Boat." ibid., February 4, 1822, 2; "A Canal Boat Adrift," ibid., March 3, 1823, 3; "For Sale, A Canal Boat," ibid., April 9, 1818, 3.

71. *Report of the Board of Public Works for 1822,* reprinted in Kohn, *Internal Improvement,* 168.

72. Porcher's *History of the Santee Canal* contains many examples of Porcher's antipathy toward Senf. Porcher stated that Senf, "seemed to be governed by an inordinate vanity which could brook not even the appearance of a rival [such as Mouzon]," and "it has been asserted, however, that it was not jealousy of Mouzon which governed Col. Senf in the choice of a location, but that he was acting in obedience to the will of one of the directors, who, possessing a large body of land on the river bank, hoped to improve the value by running the canal through it." (See Porcher, *History of the Santee Canal*). As suggested by Latrobe, the selection of the route of the Santee Canal may not have been Senf's decision. Certainly the final decision of the route of the canal would have been made by the president and directors of the Santee Canal Company. Also Porcher's criticism of the route selected for the Santee Canal might be ill founded. Subsequent engineering investigations made by the Board of Public Works and the Army Corps of Engineers do not indicate that the Greenland Swamp route would have had a better supply of water for the canal during droughts.

73. U.S. Senate, Ex. Doc. no. 60, *Letter from the Secretary of War,* 10.

74. Porcher, *History of the Santee Canal.*

75. The South Carolina legislature abolished the Board of Public Works on December 21, 1822, and replaced it with a superintendent of public works elected annually by the legislature. Blanding was elected to that position and received three thousand dollars per year.

76. Henry Glindkamp, the superintendent of the Santee Canal Company, died the following year, 1824. He had worked for the company since 1793 and was extremely popular and well respected. He was replaced by Benjamin Lequeux as superintendent of the company. On Glindkamp's death the board of directors of the company ordered the following extraordinary article published in various newspapers for a month:

> No one deserves a better tribute of respect, than he who has filled up the measure of his duty in his appointed station, and when this is one of general concern and utility, such tribute should be public. Mr. Henry Glindkamp, the late Superintendant of the Santee Canal, was at the commencement of this important and arduous enterprize, engaged in the service of the Company, and with increasing ability, perservering zeal and integrity, above suspicion, he continued to the end of his days, its active, faithful and intelligent servant. Mr. Glindkamp was the more zealous in the discharge of his duties, because he believed that the internal improvement of the State, was essentially connected with the interests committed to his care and his whole conduct was dictated and governed by a feeling and a spirit becoming a public officer. Under a deep sense of his merits in his official

and individual character, the President and Directors of the Company, do hereby

Resolve, That this Board do sincerely lament the death of their late Superintendant Henry Glindkamp, as a loss not only to his Institution, but to the Public. (*Southern Patriot,* February 26, 1824)

77. Memorandum of visit (1823).

78. *Report of the Superintendent of Public Works for 1823,* reprinted in Kohn, *Internal Improvement,* 261.

79. Ibid., 263–64.

80. Ibid., 260–63.

81. U.S. Senate, Ex. Doc. no. 60, *Letter from the Secretary of War,* 10. The company paid a 15 percent dividend in that year.

82. *Report of the Superintendent of Public Works for 1823,* reprinted in Kohn, *Internal Improvement,* 260.

83. Hollis, "Costly Delusion," 39.

84. *Report of the Superintendent of Public Works for 1827,* reprinted in Kohn, *Internal Improvement,* 517. The year is from October 1, 1826, to October 1, 1827.

85. Hollis, "Costly Delusion."

86. Ibid.

87. Ibid.

88. *Report of the Superintendent of Public Works for 1827,* reprinted in Kohn, *Internal Improvement,* 517.

89. Hollis, "Costly Delusion," 39.

90. Ibid., 39–40. The outlet to the Saluda Canal, into the Broad River, was dependent on the dam at the head of the Columbia Canal. That dam was out of commission after June 1836. See ibid., 40n66.

91. Prior to the opening of this canal there was a portage at this location.

92. Hollis, "Costly Delusion."

93. Ibid.

94. *Report of the Superintendent of Public Works for 1827,* reprinted in Kohn, *Internal Improvement,* 517. This figure of 3,144 bales shipped differs somewhat from Hollis's figure of 2,943 bales of cotton shipped. See Hollis, "Costly Delusion," 40.

95. Hollis, "Costly Delusion."

96. Ibid.

97. Ibid.

98. Ibid., 40–41.

99. *Report of the Superintendent of Public Works for 1827,* reprinted in Kohn, *Internal Improvement,* 517.

100. Hollis, "Costly Delusion," 41.

101. The numbers vary somewhat, but by 1826 it appears that 45,000 bales of cotton were being shipped through the Columbia Canal and that 199,000 bales of upcountry cotton were being shipped out of Charleston. Therefore about one quarter of the total number of bales of upcountry cotton received in Charleston transited the

Columbia Canal. See "Statement of Exports of Cotton and Rice from Charleston," *Southern Patriot,* October 2, 1827.

102. Hollis, "Costly Delusion."

103. Phillips, *A History of Transportation in the Eastern Cotton Belt,* 93–94.

104. "Internal Improvement," *Southern Patriot,* February 3, 1824, 2.

105. "A Company, Composed of Stockholders," *Southern Patriot,* May 14, 1828, 2.

106. *Report of the Board of Public Works for 1821,* reprinted in Kohn, *Internal Improvement,* 125.

107. *Report of the Superintendent of Public Works for 1826,* reprinted in ibid., 449.

108. Ibid., 449–50.

109. Ibid., 450.

110. *Report of the Board of Public Works for 1821,* reprinted in ibid., 125.

111. Ibid.

112. *Report of the Superintendent of Public Works for 1827,* reprinted in ibid., 516.

113. "Rail Road," *Southern Patriot,* March 13, 1828, 2.

114. "Internal Improvement," *Southern Patriot,* November 27, 1827, 2.

115. "Communications," *Southern Patriot,* November 29, 1827, 2.

116. Ibid.

117. "General Remarks on Rail Roads," *Southern Patriot,* December 31, 1827, 2.

118. "To the Editor," *Southern Patriot,* December 11, 1827, 2. The information in this article may be unreliable. It mentions that a committee of seven was formed but provides only five names.

119. "To the Editor of the Southern Patriot," *Southern Patriot,* December 12, 1827, 2.

120. "To the Editor," *Southern Patriot,* December 17, 1827, 2.

121. Ibid.

122. "Agreeably to a Call of the Committee," *Southern Patriot,* February 19, 1828, 2.

123. "Rail Road," *Southern Patriot,* March 13, 1828, 2.

124. "Subscription to the Railroad," *Southern Patriot,* March 15, 1828, 2.

125. "Charleston," *Southern Patriot,* March 19, 1828, 3.

126. "Rail Road Company," *Southern Patriot,* April 29, 1828, 2.

127. "We Are Gratified to Learn." *Southern Patriot,* August 20, 1828, 2.

128. Ibid. The "Engineering Department to the President" cited was most likely the U.S. Army's topographical engineers, frequently used for such civilian studies. For a study of the U.S. topographical engineers, see Schubert, *The Nation Builders.*

129. "We Are Gratified to Learn," *Southern Patriot,* August 20, 1828, 2.

130. See Fetters, *The Charleston & Hamburg,* 13–15.

131. Horatio Allen (1802–1889), civil engineer and author, was born in Schenectady, New York, and educated at Columbia College and New York University. Married to Mary Moncrief Simons, Allen received his early engineering experience on the Delaware and Hudson Canal (in Pennsylvania and New York) and the Chesapeake and Delaware Canal (in Maryland and Delaware). He was selected by the Delaware and Hudson Canal Company to travel to England to inspect and purchase steam engines. In 1829 he operated the first railroad steam engine in the United States, at Honesdale,

Pennsylvania. This was the Stourbridge Lion, imported from England. He became chief engineer of the South Carolina Canal and Rail Road Company. In this capacity he directed the construction of the railroad from Charleston to Hamburg and the construction of the first railroad steam engine sold in the United States, the *Best Friend of Charleston.* He went on to a number of other engineering positions, including assistant principal engineer on the Croton Aqueduct (New York) and chief construction engineer on the New York and Erie Railroad Company. He wrote several books, including *Astronomy in Its General Facts and Relations* (1877) and *The Railroad Era* (1884). See American Society of Civil Engineers, *A Biographical Dictionary of American Civil Engineers,* 1:2–3.

132. *Charleston Courier,* no date provided, ca. 1830, quoted by Fetters, *The Charleston & Hamburg,* 19.

133. Fetters, *The Charleston & Hamburg,* 21.

134. Ibid., 20.

135. Hollis, "Costly Delusion," 41–42. The Lockhart Canal continued in operation until 1849.

136. "Notice," *Southern Patriot,* August 24, 1827, 3.

137. Heritage Preservation Associates, "Visitor's Guide to the Old Santee Canal," 11. "Company for the Inland Navigation from Santee to Cooper River, by Canal, Petition, and Supporting Paper, for Permission to Surrender Their Charter Against Closing the Santee Canal," South Carolina Department of Archives and History S165005, item, 00050, 1853. This resulted in a counterpetition, "Residents Near Santee Canal, Counter-Petition Against Allowing the Company for the Inland Navigation from Santee to Cooper Rivers to Close the Canal, but if Allowed, That it be Transferred to Persons Willing to Keep All of it Open, and That Canal Land be Transferred Back to Original Contributors," South Carolina Department of Archives and History, S165015, item 00051, 1853.

138. Barker, *How to Kill Three Birds With One Stone.*

139. "Committee on Incorporations, Report on Various Petitions Concerning the Desire of the Company for the Inland Navigation from Santee to Cooper River to Close the Santee Canal," South Carolina Department of Archives and History, S165005, Item 00051, 1853.

140. Heritage Preservation Associates, "Visitor's Guide to the Old Santee Canal," 11.

CHAPTER 6. *Aftermath*

1. Quincy Adams Gillmore (1825–1888), military officer and civil engineer, graduated from West Point in 1849 and earned advanced degrees from Rutgers and Oberlin. He was first married to Mary Isabella O'Maher and a second time to a Mrs. Bragg. Prior to the Civil War, Gillmore worked on coastal fortifications, including those at Hampton Roads, Virginia. During the Civil War he took part of various battles, including the defense of Washington against General Jubal Early (1863). Gillmore gained a reputation as an artillerist with a capacity for breaching masonry walls using rifled cannon. Best known for his experiments and studies in cements and concrete, he published a number of important treatises on the subject, including *Practical Treatise on*

Limes (1863) and *Hydraulic Cements and Mortar* (eleven editions, 1863–96). See American Society of Civil Engineers, *A Biographical Dictionary of American Civil Engineers,* 1:50.

2. U.S. Senate, Ex. Doc. no. 60, *Letter from the Secretary of War,* 4.

3. Ibid.

4. Ibid., 5.

5. Ibid., 7.

6. Ibid.

7. Ibid., 8.

8. Ibid.

9. Ibid.

10. Ibid.

11. U. S. House of Representatives, House Joint Resolution 218, 57th Cong. 2d Sess., December 3, 1902. The resolution stated "that the Secretary of War is requested to make a survey of the country lying between the Santee and Cooper Rivers, just north of the old Santee Canal to discern the feasibility of joining the two rivers by canal. And to furnish the House of Representatives a plan and estimate of the cost of the canal by that route or by one of his choice, and to report his opinion as to the advisability to open such a canal" (Bennett and Richardson, "History of the Santee Canal," 83).

12. Bennett and Richardson, "History of the Santee Canal."

13. "Series of Maps Concerning a Proposed Change in the Santee Canal . . . 1912," a copy of which is in the South Caroliniana Library, University of South Carolina.

14. Waterman, "A Survey of the Early Buildings in the Region of the Proposed Santee and Pinopolis Reservoirs in South Carolina."

15. *Southern Patriot,* October 2, 1827, 2.

16. Board of Public Works, *Plans and Progress* (1820), reprinted in Kohn, *Internal Improvement,* 87–88.

17. Ibid., 87.

Bibliography

Allen, William C. *History of the United States Capitol: A Chronicle of Design, Construction and Politics.* Washington, D.C.: U.S. Government Printing Office, 2001.

American Society of Civil Engineers. *A Biographical Dictionary of American Civil Engineers.* 2 vols. New York: American Society of Civil Engineers, 1972, 1991.

American State Papers, Military Affairs. Vol. 3. Washington, D.C.: Gales & Seaton, 1860, report no. 362: 656–58.

Armroyd, George. *A Connected View of the Whole Internal Navigation of the United States.* Philadelphia: Published by the author, printed by Lydia R. Bailey, 1830.

Bailey, N. Louise, and Elizabeth Ivey Cooper, *Biographical Directory of the South Carolina House of Representatives,* Vol. 3, 1775–90. Columbia: University of South Carolina Press, 1980.

Bailey, N. Louise, Mary L. Morgan, and Carolyn R. Taylor, *Biographical Directory of the South Carolina Senate, 1776–1985.* 3 vols. Columbia: University of South Carolina Press, 1986.

Barker, Samuel Gaillard. *How to Kill Three Birds with One Stone: Articles Printed in the Charleston Courier.* Charleston, S.C.: A. E. Miller, 1852.

Bennett, Robert B., Jr., and Katherine H. Richardson of Heritage Preservation Associates, Inc. "History of the Santee Canal, 1785–1939." Typescript, Old Santee Canal Park, Moncks Corner, S.C., ca. 1990.

Bergeron, Jeanette, ed. *Directory of Historical Records Repositories in South Carolina.* Columbia: South Carolina State Historical Records Advisory Board, 2000.

[Blanding, Abram]. "Report of the Inspection of the Santee Canal, May 7, 1823," South Carolina Department of Archives and History, S208004, General Correspondence of the Superintendent and Board, 1820–1848, 99–103.

———. *Report of the Superintendent of Public Works on the Edisto Canal.* Columbia, S.C., October 22, 1825.

[Board of Public Works]. "Internal Improvements. Plans and Progress of Internal Improvement in South Carolina: With Observations on the Advantages Resulting Therefrom to the Agricultural and Commercial Interests of the State. Plans and Progress of the Public Works." *American Farmer* (Baltimore), December 1, 1820. Reprint of *Plans and Progress of Internal Improvement in South Carolina: With Observations on the Advantages Resulting Therefrom to the Agricultural and Commercial Interests of the State.* Charleston: W. P. Young, 1820.

Brewster, Lawrence Fay, "Ante-bellum Planters and Their Means of Transportation." In *The Proceedings of the South Carolina Historical Association,* edited by Robert D. Ochs, 15–25. Columbia: South Carolina Historical Association, 1948.

Bryan, John M. *Robert Mills: America's First Architect.* New York: Princeton Architectural Press, 2001.

Bryant, William Cullen, and Sydney Howard Gay, *A Popular History of the United States.* 4 vols. New York: Scribners, 1881–83.

Burgess, Joe B., comp. *Mount Dearborn Military Establishment.* Abbeville, S.C.: Piedmont Research Institute, 1999.

Church, Randolph W. "John Ballendine: Unsuccessful Entrepreneur of the Eighteenth Century." *Virginia Cavalcade* 8 (Spring 1959): 39–46.

Cook, James. "Letter to Mr. Timothy." *South Carolina Gazette* (Charleston, S.C.), January 18, 1770.

Crowson, E. Thomas. "Building the Landsford Canal." *South Carolina History Illustrated* 1 (May 1970): 19–22, 59–61.

———. "The Santee-Cooper: A Pioneer Effort in South Carolina Canaling." *South Carolina Magazine* 35 (August 1971): 6–20.

Department of Public Works, Superintendent, Road, River, and Canal Projects [South Carolina]. "Instructions to Walter Izard, Assistant Engineer of the Public Works." Unsigned manuscript, January 17, 1825. South Carolina Department of Archives and History.

———. "Report of the Inspection of the Santee Canal, May 7, 1823." Manuscript, South Carolina Department of Archives and History.

Diderot, Denis. *A Diderot Pictorial Encyclopedia of Trades and Industry: Manufacturing and the Technical Arts in Plates Selected from "L'Encyclopédie, ou Dictionnaire raisonné des sciences, des arts, et des métiers" of Denis Diderot.* Edited by Charles C. Gillispie. 2 vols. New York: Dover, 1993.

Drayton, Dr. Charles, Diary. Manuscript, Drayton Hall.

Drayton, John, *A View of South-Carolina, as Respects Her Natural and Civil Concerns.* Charleston, S.C.: W. P. Young, 1802. Reprint, Spartanburg, S.C.: Reprint Company, 1972.

Edgar, Walter. *History of Santee Cooper, 1934–1984.* Columbia, S.C.: R. L. Bryan, 1984.

———. *South Carolina: A History.* Columbia: University of South Carolina Press, 1998.

Farris, Don, and Nancy Farris. "The Old Canal." *Sandlapper,* Spring 2002, 56–59.

Fetters, Thomas. *The Charleston & Hamburg: A South Carolina Railroad and an American Legacy.* Charleston, S.C.: History Place, 2008.

Fleetwood, Rusty. *Tidecraft: An Introductory Look at the Boats of Lower South Carolina, Georgia, and Northeastern Florida, 1650–1950.* Savannah, Ga.: Coastal Heritage Society, 1982.

Fraser, Charles. *A Charleston Sketchbook, 1796–1806.* Charleston, S.C.: Carolina Art Association, 1940.

Gallagher, H. M. Pierce. *Robert Mills: Architect of the Washington Monument, 1781–1855.* New York: Columbia University Press, 1935.

[Gallatin, Albert]. *Report of the Secretary of the Treasury on the Subject of Public Roads and Canals; Made in Pursuance of a Resolution of the Senate, of March 2, 1807.* Washington, D.C.: R. C. Weightman, 1808. Reprint, New York: Augustus M. Kelley, 1968.

Gregg, Maxcy. "Hunting Journal." Vol. 1, 1839–1845: 6–29. Typescript. South Caroliniana Library, University of South Carolina.

Grimké, John Faucheraud. *To the Honorable the Senate and Members of the House of Representatives of the United States, the Petition and Memorial of the Subscriber, in Behalf of the Company Incorporated by Law for Opening the Catawba and Wateree Rivers.* Charleston, S.C., 1808.

Hamilton, Alexander. *The Papers of Alexander Hamilton.* Vol. 22. Edited by Harold C. Syrett. New York: Columbia University Press, 1975.

Hamlin, Talbot. *Benjamin Henry Latrobe.* New York: Oxford University Press, 1955.

Harrison, Fairfax. *Landmarks of Old Prince William: A Study of Origins in Northern Virginia.* 2 vols. Richmond: Privately printed, 1924.

Hemmings, E. Thomas. "Archeological Exploration of the Landsford Canal (38CS5)." Research Manuscript Series no. 2, April 1970. In *Notebook* (Institute of Archeology and Anthropology, University of South Carolina, Columbia), edited by Robert L. Stephenson. Vol. 4 (March–April 1972): 33–56.

Heritage Preservation Associates. "Visitor's Guide to the Old Santee Canal: A Short History." Typescript, Old Santee Canal Park, Moncks Corner, S.C., 1989.

Holley, Byron. *Prospectus of the Enlarged Columbia Canal, Columbia, S.C. with a Map of Location, Prepared for Publication Under the Direction of the Board of Trustees.* Charleston, S.C.: Printed by Walker, Evans & Cogswell, 1891.

Hollis, Daniel W. "Costly Delusion: Inland Navigation in the South Carolina Piedmont." In *The Proceedings of the South Carolina Historical Association,* 29–43. Columbia: South Carolina Historical Association, 1968.

"Internal Improvements in South Carolina." *North American Review* 4 (July 1821): 143–54.

Jackson, Melvin H. *Privateers in Charleston, 1793–1796: An Account of a French Palatinate in South Carolina.* Washington, D.C.: Smithsonian Institution Press, 1969.

Jenkins, John S. *Daring Deeds of American Generals.* New York: A. A. Kelley, 1856.

Kapsch, Robert J. *Canals.* Norton / Library of Congress Visual Sourcebook in Architecture, Design, and Engineering. New York: Norton, 2004.

———. "The Conewago Canal: Pennsylvania's First Canal." In *Canal History and Technology Proceedings,* edited by Lance E. Metz, 23:5–43. Easton, Pa.: Canal History and Technology Press, National Canal Museum, 2004.

———. "George Washington, the Potomac Canal and the Beginning of American Civil Engineering History: Problems and Solutions." In *American Civil Engineering History: The Pioneering Years,* edited by Bernard G. Dennis Jr., Robert J. Kapsch, Robert LoConte, Bruce W. Mattheiss, and Steven M. Pennington, 129–94. Reston, Va.: American Society of Civil Engineers, 2003.

———. "The Potomac Canal: A Construction History," In *Canal History and Technology Proceedings,* edited by Lance E. Metz, 21:143–235. Easton, Pa.: Canal History and Technology Press, National Canal Museum, 2002.

———. *The Potomac Canal: George Washington and the Waterway West.* Morgantown: West Virginia University Press, 2007.

———. "The Santee Canal: America's First Summit Level Canal." In *Canal History and Technology Proceedings,* edited by Lance E. Metz, 24:22–80. Easton, Pa.: Canal History and Technology Press, 2005.

Kapsch, Robert J., and Elizabeth Perry Kapsch, *The Monocacy Aqueduct on the Chesapeake and Ohio Canal.* Poolesville, Md.: Medley Press / Center for Historic Engineering and Architecture Research, 2005.

Kapsch, Robert J., and Yvonne Long. "James Brindley, American Canal Engineer." *Transactions of the Newcomen Society* (forthcoming 2010).

Kirby, Richard Shelton. "William Weston and His Contribution to Early American Engineering." *Transactions of the Newcomen Society* 16 (1935–36): 1–17.

Knight, Edward H. *Knight's American Mechanical Dictionary.* 3 vols. New York: Hurd & Houghton, 1877.

Kohn, David, and Bess Glenn, eds. *Internal Improvement in South Carolina 1817–1828 from the Reports of the Superintendent of Public Works and from Contemporary Pamphlets, Newspaper Clippings, Letters, Petitions and Maps.* Washington, D.C. Privately printed, 1938. Includes: *Report of the Civil and Military Engineer of the State of South-Carolina for the Year 1818; Report No. 1 of the Civil and Military Engineer of the State of South-Carolina for the Year 1819;* Board of Public Works, *Plans and Progress of Internal Improvement in South Carolina: With Observations on the Advantages Resulting Therefrom to the Agricultural and Commercial Interests of the State* (Charleston: W. P. Young, 1820); Robert Mills, *A Treatise on Inland Navigation Accompanied by a Map* [excerpt] (Baltimore: F. Lucas Jr. 1829); *Report of the Board of Public Works, to the Legislature of South Carolina, for the Year, 1821;* Robert Mills, *Inland Navigation. Plan for a Great Canal between Charleston and Columbia, and for Connecting Our Waters with Those of the Western Country* [excerpt] (Columbia: Telescope Press, 1821); *Report of the Board of Public Works, to the Legislature of South Carolina, for the Year, 1822;* Robert Mills, *Internal Improvement of South-Carolina, Particularly Adapted to the Low Country* (Columbia: State Gazette Office, 1822); *Report of the Superintendent of Public Works, to the Legislature of South Carolina. for the Year 1823* (Columbia: D. & J. M. Faust, 1823); *The Annual Report of the Superintendent of Public Works of South Carolina, for 1824* (Columbia: Black & Sweeny, 1824); *The Report of the Superintendent of Public Works of South Carolina, for 1825* (n.d.); *The Report of the Superintendent of Public Works, for 1826* (Columbia: D. E. Sweeny, 1826); *The Report of the Superintendent of Public Works of South Carolina, to the Legislature of South Carolina, for the Year 1827* (Columbia: Sweeny & Sims at the Telescope Press, 1827); *Report of the Superintendent of Public Works, for the Year 1828* (n.d.).

Langley, Mary L., and Albert M. Langley. *The Railroad Comes of Age: The Historic South Carolina Canal and Railroad and the Best Friend of Charleston.* Augusta, Ga.: Augusta Chapter, National Railway Historical Society, 1970.

Latrobe, Benjamin. *The Correspondence and Miscellaneous Papers of Benjamin Henry Latrobe.* Edited by John C. Van Horne, Jeffrey A. Cohen, Darwin H. Stapleton,

William B. Forbush III, and Tina H. Sheller. 3 vols. New Haven: Yale University Press for the Maryland Historical Society, 1988.
Leckie, Robert. Papers of Robert Leckie (ca. 1825 to July 20, 1834). Perkins Library, Duke University Library.
Levine, Ida L. "A Letter from William Moultrie at Charleston to George Washington at Mount Vernon [Concerning the Santee Canal and the Use of James Brindley], April 7, 1786," *South Carolina Historical Magazine* 83 (April 1982): 116–19.
Linklater, Andro. *Measuring America: How the United States Was Shaped by the Greatest Land Sale in History.* New York & London: Penguin, 2003.
Lossing, Benson J. *Pictorial Fieldbook of the American Revolution.* 2 vols. New York: Harper, 1859.
McCord, David J., ed. *The Statutes at Large of South Carolina: Edited under Authority of the Legislature.* Vols. 6 and 7. Columbia, S.C.: A. S. Johnston, 1840.
Meyer, Balthasar Henry, Caroline E. MacGill, et al. *History of Transportation in the United States before 1860.* Publication no. 215c. Washington, D.C.: Carnegie Institution of Washington, 1917. Reprint, Forge Village, Mass.: Peter Smith, 1948.
Mills, Robert. *Atlas of the State of South Carolina.* Columbia, S.C., 1825. Reprinted as *Mills' Atlas: Atlas of the State of South Carolina, 1825.* Easley, S.C.: Southern Historical Press, 1980.
———. *Inland Navigation. Plan for a Great Canal between Charleston and Columbia and for Connecting Our Waters with Those of the Western Country.* Columbia, S.C.: Telescope Press, 1821.
———. *Internal Improvement of South-Carolina, Particularly Adapted to the Low Country.* Columbia, S.C.: State Gazette Office, 1822.
———. Papers of Robert Mills, 1781–1855. Library of Congress. Published in microform as *The Scholarly Resources Microfilm Edition of the Papers of Robert Mills, 1781–1855,* edited by Pamela Scott. Wilmington, Del.: Scholarly Resources, 1990.
———. *Statistics of South Carolina, Including a View of its Natural, Civil, and Military History, General and Particular.* Charleston, S.C.: Hurlbut & Lloyd, 1826. Spartanburg, S.C.: Reprint Company, 1972.
———. *A Treatise on Inland Navigation.* Baltimore: F. Lucas Jr., 1820.
Mitchell, S. Augustus. *Mitchell's Compendium of the Internal Improvement of the United States.* Philadelphia: Mitchell & Hinman, 1835.
Mouzon, Henry, Jr. *An Accurate Map of North and South Carolina, with Their Indian Frontier.* London: Printed for Robt. Sayer & J. Bennett, 1775.
[———]. "A Map of the Parish of St. Stephen, in Craven County, Establishing a View of the Several Places Practicable for Making a Navigable Canal between Santee and Cooper Rivers," ca. 1785. Ascribed to Henry Mouzon Jr. by the South Caroliniana Library. One version of this map is reproduced in Porcher, *The History of The Santee Canal,* and in U.S. Senate, Ex. Doc. no. 60, 46th Cong., 3d sess., February 28, 1881, *Letter from the Secretary of War Transmitting a Communication from the Chief of Engineers of February 25, and Accompanying a Report from Lieut. Col. Q. A. Gillmore, Corps of Engineers.* This map is hand dated "1775" in Porcher's history, but that date may

have been added later. The South Caroliniana Library assigns the date "1785" to its copy, which is similar but not identical to Porcher's.

Newell, Mark M. "Buried Treasure: An Archeologist Finds the Beginnings of America's Canal Age Mired in a South Carolina Swamp." *Invention and Technology* 8 (Winter 1992): 35–41.

Newell, Mark M., and Joe J. Simmons III. *The Santee Canal Sanctuary.* Part 2, *Preliminary Archaeological Investigation of a Portion of the Old Santee Canal and Biggin Creek, Berkeley County, South Carolina.* Columbia: Underwater Antiquities Management Program of the South Carolina Institute of Archaeology and Anthropology, University of South Carolina, February 1989.

Phillips, Ulrich Bonnell. *A History of Transportation in the Eastern Cotton Belt to 1860.* New York: Columbia University Press, 1908. Reprint, New York: Octagon Books, 1968.

Placzek, Adolf K., ed. *Macmillan Encyclopedia of Architects.* 4 vols. New York: Free Press, 1982.

Porcher, Frederick A. *The History of the Santee Canal, Prepared by the Late Prof. F. A. Porcher and Dedicated to the South Carolina Historical Society, 1875.* With an appendix by A. S. Salley Jr. Charleston: South Carolina Historical Society, 1903. Reprint, Moncks Corner: South Carolina Public Service Authority, 1950.

Preservation Consultants, Inc., with Robert P. Stockton. *Historic Resources of Berkeley County, South Carolina.* South Carolina: Berkeley County Historical Society / South Carolina Department of Archives and History, 1990.

Pruitt, A. B. *Creeks and Rivers in South Carolina.* N.p.: Privately published, 2004.

Raiford, Norman Gasque. "South Carolina and the Issue of Internal Improvement, 1775–1860." Ph.D. diss., University of Virginia, 1974.

Richardson, Lewis W. "The Canals of South Carolina." In *The Best from American Canals.* Vol. 1: 34–35. York, Pa.: American Canal and Transportation Center, 1980.

Ristow, Walter. "Robert Mills' Atlas of South Carolina, 1825: The First American State Atlas." *Quarterly Journal of the Library of Congress* 34 (January 1977): 52–66.

Rogers, George C., Jr. *Evolution of a Federalist: William Loughton Smith of Charleston (1758–1812).* Columbia: University of South Carolina Press, 1962.

———, ed. "The Letters of William Loughton Smith to Edward Rutledge (1789–94)," part 3. *South Carolina Historical Magazine* 69 (October 1968): 225–42.

Salley, A. S., ed. *History of the Old Santee Canal.* Moncks Corner: South Carolina Public Service Authority, ca. 1944.

[Sampit and St. James Santee Canal Company]. *Rules and Regulations of the Sampit and St. James Santee Canal Company, Passed December 19, 1809.* Charleston, S.C.: Gabriel M. Bounetheau, 1810.

[Santee Canal Company]. *The By-laws of the Company for the Inland Navigation from Santee to Cooper River (Commonly Called the Santee Canal Company), Adopted by the Stockholders at a Meeting Held on the 23d January, 1841, to Which is Prefixed the Act of the Legislature Passed on 22d March, 1786, for Incorporating the Company.* Charleston, S.C.: A. E. Miller, 1841.

———. [Petition for the Santee Canal Company], February 7, 1786. South Carolina Department of Archives and History, series S165015, 1786, no. 5.

———. "Report of Board of Directors," 1805. South Carolina Historical Society.

Schubert, Frank N., ed. *The Nation Builders: A Sesquicentennial History of the Topographical Engineers.* Office of History, United States Corps of Engineers, Fort Belvoir, Virginia (Washington, D.C.: U.S. Government Printing Office, 1988).

Scott, Pamela, ed. *Guide and Index to the Scholarly Resources Microfilm Edition of The Papers of Robert Mills 1781–1855.* Wilmington, Del.: Scholarly Resources, 1990.

Senf, John Christian. *General Plan of the Canal and Its Environs between Santee and Cooper Rivers in the State of South Carolina, Commenced in the Year 1793 and Finished in the Year 1800 by Christian Senf, Colonel, Engineer and Director in Chief of the Canal,* redrawn by J. P. Gaillard. N.p., n.d.

Senf, John Christian, Santee Canal Company (S.C.). "Report to the President and Board of Directors, 1800 November 11." Manuscript (in Senf's handwriting), South Carolina Historical Society. Published in Army Report of 1881, 11–25, and in Mabel L. Webber, ed., "Col. Senf's Account of the Santee Canal," *South Carolina Historical and Genealogical Magazine* 28 (January 1927): 8–21, 112–31.

Shank, William H., "John Christian Senf." In *Towpaths to Tugboats: A History of American Canal Engineering,* 15–16. York, Pa.: American Canal and Transportation Center, 1982.

Simmons, Joe J., III, and Mark M. Newell. *The Santee Canal Sanctuary.* Part 1, *Preliminary Archaeological Surveys of a Portion of the Old Santee Canal, the Biggin Canal Vessel and the Mouth of Biggin Creek, Berkeley County, South Carolina.* Columbia: Underwater Antiquities Management Program of the South Carolina Institute of Archaeology and Anthropology, University of South Carolina, February 1989.

Skaggs, David Curtis. "John Semple and the Development of the Potomac Valley, 1750–1773." *Virginia Magazine of History and Biography* 92 (July 1984): 282–308.

Skempton, A. W., M. M. Chrimes, R. C. Cox, P. S. M. Cross-Rudkin, R. W. Rennison, and E. C. Ruddock, eds. *A Biographical Dictionary of Civil Engineers in Great Britain and Ireland.* Vol. 1, *1500–1830.* London: Thomas Telford for the Institution of Civil Engineers, 2002.

Sobel, Robert, and John Raimo, eds. *Biographical Directory of the Governors of the United States, 1789–1978.* Reprint edition, Westport, Conn.: Meckler, 1978.

South Carolina Department of Parks, Recreation and Tourism. *South Carolina River Trails.* N.p., 1978.

South Carolina legislature. *An Act to Establish a Company for Clearing and Improving the Navigation of the Edisto and Ashley Rivers, and the Forming a Communication by a Canal and Locks between the Former and the Latter,* 1787.

———. *An Act to Establish a Company for the Opening of the Navigation of the Catawba and Wateree Rivers,* 1787.

South Carolina Public Service Authority. *History of the Old Santee Canal and How This 18th Century Waterway of Our Founding Fathers Lives Again in the Santee-Cooper Hydro-Electric and Navigation System of Today.* Moncks Corner: South Carolina Public Service Authority, n.d.

[Susquehanna Canal Co.]. *Report of the Governor and Directors to the Proprietors of the Susquehanna Canal.* N.p., 1802.

Tanner, Henry S. *A Description of the Canals and Rail Roads of the United States.* New York: Tanner & J. Disturnell, 1840.

Taylor, George Roberts. *The Transportation Revolution, 1815–1860.* New York: Rinehart, 1951. Reprint, New York: Harper & Row, 1968.

U.S. Congress. *Report of the Committee on Roads and Canals, on the Subject of Internal Improvements, Accompanied with a Bill to Procure the Necessary Surveys, etc. on the Subject of Roads and Canals.* Washington, D.C.: 1822.

———. *Report of the Committee to Whom Was Referred on the Fourteenth November Last, the Petition of the Company for Opening the Navigation of the Catawba and Wateree Rivers,* January 26, 1809. Washington, D.C.: Printed by A. & C. Way, 1809.

U.S. Senate, Ex. Doc. no. 60. *Letter from the Secretary of War Transmitting a Communication from the Chief of Engineers of February 25, and Accompanying a Report from Lieut. Col. Q. A. Gillmore, Corps of Engineers, upon Survey for the Reopening of the Santee Canal, South Carolina, etc.,* 46th Cong., 3d sess., February 28, 1881.

Von Gerstner, Franz Anton Ritter. *Die Innern Communicationen der Vereinigten Staaten von Nordamerica.* 1842–43. Translated by David J. Diephouse and John C. Decker as *Early American Railroads: Franz Anton Ritter von Gerstner's Die innern Communicationen (1842–1843),* edited by Frederick C. Gamst. Stanford, Calif.: Stanford University Press, 1997.

Wade, Arthur P. "Mount Dearborn: The National Armory at Rocky Mount, South Carolina, 1802–1829. *South Carolina Historical Magazine* 81 (July 1980): 207–31; 81 (October 1980): 316–41.

Wallston, R. E., and Joel R. Poinsett. Papers, Edisto and Ashley River Canal Company, 1801–9. South Carolina Historical Society.

Washington, George. *The Diaries of George Washington 1748–1799.* Edited by John C. Fitzpatrick. 4 vols. Boston & New York: Houghton Mifflin, 1925.

———. *The Papers of George Washington,* Confederation Series. Edited by W. W. Abbott, Dorothy Twohig, et al. Vols. 3–5. Charlottesville: University Press of Virginia, 1995.

Waterman, Thomas T. "A Survey of the Early Buildings in the Region of the Proposed Santee and Pinopolis Reservoirs in South Carolina." Typescript. Washington, D.C., Historic American Buildings Survey, U.S. Department of the Interior, 1939. Library of Congress.

Weaver, Charles Clinton. *Internal Improvements in North Carolina Previous to 1860,* Johns Hopkins University Studies in Historical and Political Science, series 21, no. 3–4. Baltimore: Johns Hopkins Press, March–April 1903.

Webber, Mabel L., ed., "Col. Senf's Account of the Santee Canal." *South Carolina Historical and Genealogical Magazine* (South Carolina Historical Society) 28 (January 1927): 8–21; 112–31.

Workers of the Writers' Program of the Works Projects Administration. *South Carolina: A Guide to the Palmetto State.* New York: Oxford University Press, 1941.

Wright, Louis B. *South Carolina: A Bicentennial History.* New York: Norton, 1976.

Index

Page numbers in italics refer to illustrations.

About the Author

ROBERT J. KAPSCH is a researcher and writer for the Center for Historic Engineering and Architecture. Previously he served as senior scholar in historic architecture and engineering for the National Park Service, as project engineer for the Chesapeake and Ohio Canal National Historical Park, and as chief of the Historic American Buildings Survey / Historic American Engineering Record. Kapsch was awarded the Distinguished Service Medal, the highest award of the U.S. Department of Interior, for his work at the National Park Service. He is the author of *Canals* and *The Potomac Canal: George Washington and the Waterway West* (Silver Medal Winner, 2008 *ForeWord* Book of the Year for History) and coauthor with his wife, Elizabeth Perry Kapsch, of *The Monocacy Aqueduct on the Chesapeake and Ohio Canal* and *Seneca, Maryland: Standing Stones.* He lives in North Potomac, Maryland.